Comprehensive Techniques in CSF Leak Repair and Skull Base Reconstruction

Advances in Oto-Rhino-Laryngology

VOL. 74

Series Editors

G. Randolph Boston, Mass.

Comprehensive Techniques in CSF Leak Repair and Skull Base Reconstruction

Volume Editor

Benjamin S. Bleier Boston, Mass.

74 figures, 49 in color, and 12 tables, online supplementary material, 2013

KARGER

Basel · Freiburg · Paris · London · New York · New Delhi · Bangkok · Beijing · Tokyo · Kuala Lumpur · Singapore · Sydney

Benjamin S. Bleier
Harvard Medical School
Massachusetts Eye and Ear Infirmary
Department of Otolaryngology, Head and Neck Surgery
243 Charles Street
Boston, MA 02114 (USA)

Library of Congress Cataloging-in-Publication Data

Comprehensive techniques in CSF leak repair and skull base reconstruction / volume editor, Benjamin S. Bleier.
p. ; cm. -- (Advances in oto-rhino-laryngology, ISSN 0065-3071 ; vol. 74)
Includes bibliographical references and indexes.
ISBN 978-3-8055-9952-8 (hardcover : alk. paper) -- ISBN 978-3-8055-9953-5 (e-ISBN)
I. Bleier, Benjamin S. II. Series: Advances in oto-rhino-laryngology ; v. 74. 0065-3071
[DNLM: 1. Cerebrospinal Fluid Rhinorrhea--surgery. 2. Postoperative Complications--prevention & control. 3. Reconstructive Surgical Procedures--methods. 4. Skull Base--surgery. 5. Surgical Flaps. W1 AD701 v. 74 2013 / WL 354]

617.5'14--dc23

2012037119

Bibliographic Indices. This publication is listed in bibliographic services, including Current Contents®.

Disclaimer. The statements, opinions and data contained in this publication are solely those of the individual authors and contributors and not of the publisher and the editor(s). The appearance of advertisements in the book is not a warranty, endorsement, or approval of the products or services advertised or of their effectiveness, quality or safety. The publisher and the editor(s) disclaim responsibility for any injury to persons or property resulting from any ideas, methods, instructions or products referred to in the content or advertisements.

Drug Dosage. The authors and the publisher have exerted every effort to ensure that drug selection and dosage set forth in this text are in accord with current recommendations and practice at the time of publication. However, in view of ongoing research, changes in government regulations, and the constant flow of information relating to drug therapy and drug reactions, the reader is urged to check the package insert for each drug for any change in indications and dosage and for added warnings and precautions. This is particularly important when the recommended agent is a new and/or infrequently employed drug.

www.karger.com
Printed in Germany on acid-free and non-aging paper (ISO 97069) by Kraft Druck, Ettlingen
ISSN 0065–3071
e-ISSN 1662–2847
ISBN 978–3–8055–9952–8
e-ISBN 978–3–8055–9953–5

I would like thank both my mentors and students from whom
I continue to learn how to better care for our patients. Most importantly,
I dedicate this book to my wife whose love and support form the foundation
upon which any of my accomplishments rest.

Contents

OSM

Online supplementary material: www.karger.com/adorl074_suppl

Preface

The discipline of endoscopic skull base surgery has evolved profoundly in the past several decades fueled largely by advances in imaging, instrumentation and endoscopic techniques. The collaboration of neurosurgical and otolaryngologic surgeons has led to the elaboration of a series of endoscopic corridors that may be used to approach and resect lesions involving multiple subsites along the cranial base. As the indications for these approaches continue to expand, the ability to reconstruct the skull base in an immunocompetent and watertight fashion has emerged as a critical rate-limiting factor. Borrowing from traditional methods in open reconstruction and neurotologic surgery, early reconstructive approaches relied on multilayer free grafts. As defect size increased however, the failure rates associated with avascular tissue repairs provided the impetus for the development of a variety of intranasal and extranasal flap techniques. These methods have been remarkably successful resulting in dramatic reductions in postoperative cerebrospinal fluid leaks and overall morbidity.

This textbook represents a detailed exposition of the techniques required for successful skull base reconstruction written by many of the world experts who have played an important role in the expansion of this rapidly evolving field. Through textual descriptions, figures, and endoscopic videos, readers of all levels of endoscopic expertise will be able to use this book to improve and refine their techniques.

While this text represents the current state of the art in cerebrospinal fluid leak and skull base repair, the final chapter covers new technologies which are beginning to appear on the horizon and may continue to push the boundaries of what is possible using current approaches.

Endoscopic skull base surgery has allowed for the management of complex lesions in a minimally invasive fashion resulting in improved quality of life and decreased hospital stays. Advances in techniques of cranial base reconstruction have been crucial to the progression of this field. This textbook is written as a practical guide to enable the skull base surgeon to understand all the reconstructive options currently available and to maximize success in the management of these complex patients.

Benjamin S. Bleier, MD, Boston, Mass., USA

Bleier BS (ed): Comprehensive Techniques in CSF Leak Repair and Skull Base Reconstruction.
Adv Otorhinolaryngol. Basel, Karger, 2013, vol 74, pp 1–11 (DOI: 10.1159/000342263)

Skull Base Anatomy and CSF Rhinorrhea

Allen Lam · Eric Holbrook

Massachusetts Eye and Ear Infirmary, Boston, Mass., USA

Abstract

The skull base is an intricate interface between the cranium and face allowing the passageway of vital structures. Anatomic conceptualization forms the platform for surgical approaches and solutions to skull base pathology. Understanding embryogenesis provides further depth into the pathophysiology of congenital defects. This chapter aims to highlight skull base anatomy and embryology along with cerebrospinal fluid physiology and diagnostic evaluation of cerebrospinal fluid leaks.

Skull Base Anatomy

The skull base defines the inferior aspect of the skull that divides the neurocranially derived cranial contents from the viscerocranial structures of the face and neck. Cranial nerves and vital vasculature travel through the skull base via foramina. Given the complexity of the skull base and the importance of these structures, in-depth anatomic knowledge is critical for operating in this space. The skull base is divided into 3 distinct fossae: anterior, middle and posterior.

Anterior Cranial Fossa

The anterior cranial fossa is formed by the frontal, sphenoid, and ethmoid bones. Anteriorly, the frontal bone separates fossa from the frontal sinus while the floor partitions the orbital contents and the nasal cavity. The midline frontal crest divides the cerebral hemispheres and attaches the falx cerebri. The cribriform plate and crista galla lie more posteriorly allowing passage of the olfactory nerves. It lies in a depressed plane between the orbits as part of the nasal cavity roof, an overhang easily violated. The lateral lamellae of the cribriform plate attach superiorly to the orbital plates of the frontal bone via the fovea ethmoidalis creating the roof of the ethmoid air cells. The Keros classification system stratifies the height of the cribriform plate into 3 categories, characterizing the risk of injury to the overhanging skull base [1].

The anterior ethmoidal vessels and nerve arc across the roof of the nasal cavity at the intersection of the fovea ethmoidalis with the lateral lamella. The posterior ethmoidal nerve and vessels run near the posterolateral corner of the cribriform plate. The planum sphenoidale lies in continuum with the lesser wings of the sphenoid denoting the posteromedial limit of the anterior cranial fossa (fig. 1) [2, 3].

Middle Cranial Fossa

The middle cranial fossa is formed primarily by the sphenoid bone inclusive of its body and the greater and lesser wings. The greater wings fan out laterally to the calvarial wall articulating with the frontal, temporal and parietal bones. The lesser wings span the superior margin of the superior orbital fissure and form an attachment for the tentorium cerebelli. Inferiorly, the pterygoid plates drop down at the intersection of the sphenoid body and greater wings [3].

The body houses the sphenoid sinus and sella turcica, seating the pituitary gland. The chiasmatic sulcus provides a groove for the optic nerves and drops off just posterior to the planum sphenoidale anterior to the sella. The concave sides of the sphenoid body are sculpted by the sigmoid portion of the internal carotid artery (ICA) while the clivus downslopes to the basiocciput. Inferiorly, the body forms the roof of the nasopharynx and articulates medially with the vomer via the sphenoid rostrum. The medial pterygoid plates drop from the sphenoid body to form the lateral limits of the nasopharynx [3].

The orbit lies inferior to the anterior cranial fossa and anterior to the middle cranial fossa housed by components of the frontal, sphenoid, zygomatic, lacrimal, nasal and ethmoid bones. The anterior and posterior ethmoid artery foramina lie 12 and 24 mm behind the anterior lacrimal crest, respectively, with the optic canal lying 6 mm behind the posterior ethmoid foramina [4]. The superior orbital fissure is comprised of a medial compartment transmitting the oculomotor, nasociliary and abducens nerve along with sympathetic innervation and the superior ophthalmic veins, and a lateral compartment that transmits the lacrimal, frontal and trochlear nerves. The optic canal carries the optic nerve and the ophthalmic artery. In continuum with the superior orbital fissure and optic canal lies the inferior orbital fissure which relays the infraorbital and zygomatic nerves, the infraorbital artery and vein, and the inferior ophthalmic vein.

Important foramina form passages into and out of the middle cranial fossa. Foramen rotudum and the vidian (pterygoid) canal pass inferior to the superior orbital fissure. Foramen rotundum transmits the maxillary branch of the trigeminal nerve. The vidian canal lies inferomedial to foramen rotundum and can be used to identify the petrous portion of the ICA. The palatovaginal, or palatosphenoid, canal lies medial to the vidian canal conveying the pharyngeal branch of the maxillary artery and branches of the pterygopalatine ganglion. Endoscopically, the palatovaginal canal can help identify the vidian canal, thereby protecting the vidian nerve and ICA [5]. Foramen ovale and spinosum lie posterolateral to the rotundum at the

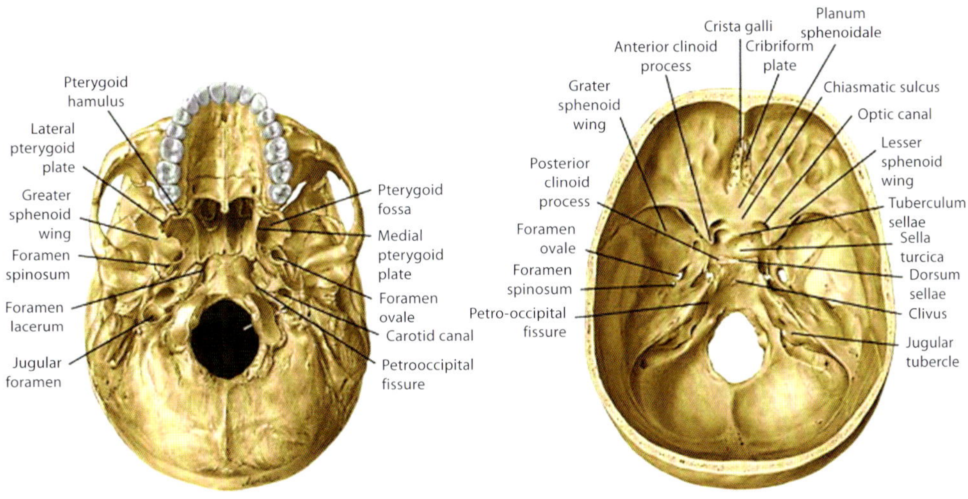

Fig. 1. Cranial and caudal view of the skull base.

intersection of the greater wing and body of the sphenoid. Foramen ovale carries the mandibular branch of the trigeminal nerve and spinosum, the middle meningeal artery. The Gasserian ganglion of the trigeminal nerve sits in the depression of the petrous apex known as Meckel's cave. The greater superficial petrosal nerve courses into foramen lacerum joining the deep petrosal nerve to become the nerve of the pterygoid canal or vidian nerve. Followed posteriorly, this nerve and canal project to the ICA, an important landmark in skull base surgery. The lesser superficial petrosal nerve lies parallel to the greater superficial petrosal nerve and exits via foramen ovale towards the otic ganglion [6].

The cavernous sinus lies within the dura of the middle cranial fossa adjacent to the sphenoid body. The ICA courses through the cavernous sinus in an S-shaped path and delivers 2–6 caroticocavernous branches which supply the pituitary. The abducens nerve runs along the lateral wall of the ICA as it enters from Dorello's canal. The oculomotor and trochlear nerves also run within the dura but not within the cavernous sinus itself.

Posterior Cranial Fossa

The posterior aspect of the petrous portion of the temporal bone creates the anterolateral wall of the posterior cranial fossa. The clivus, a synostosis of the basisphenoid and basiocciput, creates the anteromedial portion. The floor and posterolateral aspects are made up of the occipital bone while the tentorum cerebelli makes up the roof.

Five foramina convey neurovascular structures to and from the posterior cranial fossa. The internal acoustic canal and vestibular aqueduct convey the facial and vestibulocochlear nerves along with nervous intermedial and branches of the anterior inferior cerebellar artery. The jugular foramen contains the glossopharyngeal, vagus, and spinal accessory nerves along with the confluence of the sigmoid sinus, inferior petrosal sinus, and jugular bulb as they descend to become the internal jugular vein. The hypoglossal canal carries the hypoglossal nerve and a meningeal branch of the ascending pharyngeal artery. Finally, foramen magnum transmits the medulla oblongata, the spinal accessory nerve and spinal branches, and the vertebral arteries.

Skull Base Embryology

The skull base develops through a complex interaction of cellular proliferation, migration, resorption and growth. Only after formation of the primitive brain and cranial nerves does the skull base begin to form [3]. Condensations of neural crest cells form the desmocranium which become multiple chondrification centers around the 7th week of development. Prechordal and parachordal cartilages ossify to form the basiocciput and basisphenoid while presphenoid condensations create the anterior sphenoid. Mesoethmoid cartilages develop into the perpendicular plate of the ethmoid bone and crista galli. Orbitosphenoid cartilages become the lesser wings of the sphenoid while alisphenoid cartilages create the greater wings. Cartilages fuse around pre-existing structures to form the cranial base foramina [7]. Both intramembranous and endochondral ossification contribute to the development of the bony skull base [3].

Congenital Defects of the Skull Base

Incomplete fusion of the ossification centers weakens the integrity of the skull base. CSF fistulae may occur at these points presenting as CSF rhinorrhea and/or encephalocele herniation. Anterior cranial fossa defects result from incomplete regression of the midline dural diverticulum, foramen cecum. The fonticulus frontalis transiently lies between the developing frontal and nasal bones through which frontonasal encephaloceles herniate. Similarly, frontoethmoidal encephaloceles herniate via the foramen cecum through defects in the ethmoid ala of the cribriform plate. About 15% of encephaloceles are frontal, presenting as midline nasal masses that enlarge with valsava maneuvering or jugular compression (Furstenberg sign) [8].

Multiple defects may arise from the sphenoid sinus due to its complex formation with coalescences of the alisphenoid, basisphenoid, and presphenoid ossification centers. The lateral craniopharyngeal canal was described by Sternberg [9] in 1888 as a defect linking the sphenoid body to the lateral wall of the sphenoid sinus. Encephaloceles through Sternberg's canal are thought to be related to overpneumatization of the sphenoid sinus [10, 11]. An enphalocele herniating through the right lateral sphenoid recess is shown in figure 2. However, the existence of

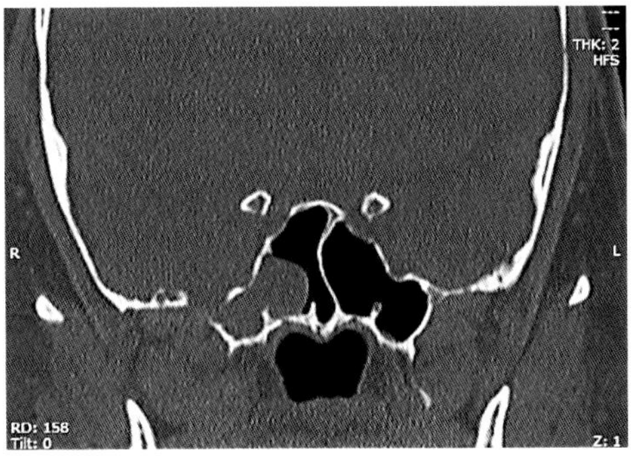

Fig. 2. High-resolution CT coronal reformat demonstrating a right lateral sphenoid recess skull base defect with encephalocele.

Sternberg's canal is contentious due to inconsistencies of where most lateral sphenoid sinus fistulae occur, most described occurring lateral to V2 [12]. Therefore, for this theory to be true, the encephalocele would need to pass through the cavernous sinus and two layers of dura [12].

Lateral skull base defects such as a patent Hyrtl fissure or widened petromastoid suture may manifest as CSF otorrhea. Though interesting, these are out of the scope of this chapter but should be considered when evaluating patients with CSF rhinorrhea as the eustachian tubes may convey CSF.

CSF Physiology

CSF is a clear, colorless fluid which performs multiple functions including a protective cushion and a transport medium for metabolites, toxins, and neurohormonal signals. CSF flows from the lateral ventricles to the third ventricle via the foramen of Monro then into the fourth ventricle via the aqueduct of Sylvius then into the central canal of the spinal cord. CSF enters the subarachnoid space from the fourth ventricle via the foramen of Magendie and the paired foramen of Luschka to circulate around the spinal cord, cerebellum and cerebrum. Ninety percent of the circulating CSF is absorbed by the arachnoid venules feeding into the dural venous sinuses, while the remaining 10% exits by cerebral venules [13, 14].

Fifty percent of CSF is produced by the ependyma of the choroid plexus lining the third, fourth, and lateral ventricles. The remaining 50% arises from interstitial fluid shifts from the ventricular walls. Compared to plasma, CSF has equivalent sodium concentrations, higher chloride concentrations, and lower concentrations of protein, glucose, and potassium. The total volume of CSF is approximately 150 ml with adult production being 550 ml/day, or 20 ml/h. Normal intracranial pressures range from 5 to 15 mm Hg. While production of CSF remains constant, absorption of CSF increases directly proportional to increases in pressure. At 11 mm Hg, the rate of secretion equals the rate of absorption [15] (table 1).

Table 1. Composition of lumbar CSF compared to blood

Constituent	Lumbar CSF	Blood
Na^+, mEq/l	148	136–145
K^+, mEq/l	2.9	3.5–5
Cl^-, mEq/l	120–130	100–106
Glucose, mg/dl	50–75	70–100
Protein, mg/dl	15–45	6.8×10^3
pH	7.3	7.4

CSF Leak Diagnosis

Diagnosis of a CSF leak can present a clinical challenge, especially if the leak is spontaneous or idiopathic in nature. CSF leaks have been well documented as far back as the 2nd century when Galen described leakage of CSF following head trauma [16]. The term 'rhinorrhea' was coined in 1899 by St Clair Thompson describing a cohort of patients with spontaneous CSF leak [17]. However, it was not until the 20th century when Dandy [18] first reported successful repair of a CSF leak with fascia lata. Currently, a proper history and clinical evaluation still provides important insight into potential etiology, location, and amount of CSF leak as well as complications. Fluid sample analysis and radiographic imaging can yield additional information. Detection and localization of the defect are critical for successful repair of the CSF leak site.

History and Physical Examination

CSF rhinorrhea usually presents as clear nasal drainage. Patients may complain of headaches and other neurologic symptoms such as change in vision, tinnitus, dizziness or paresthesias. Symptoms may worsen with valsalva maneuvers or with positioning such as leaning forward or lying supine. Hyposmia or anosmia may be present if the cribriform plate is involved. More concerning are patients with symptoms of meningitis such as neck stiffness, photophobia and persistent headache. As the major cause of morbidity and mortality, meningitis has been reported to complicate untreated CSF leaks in approximately 25–50% of cases with risk increasing with duration of CSF leak [19, 20]. Antecedent events such as blunt trauma or recent sinonasal, otologic or neurologic surgery are also important to ascertain.

On physical examination, quality, quantity and laterality of rhinorrhea should be assessed. Having the patient bend forward may elicit a gush of rhinorrhea, and the fluid should be sent for analysis. The halo sign has historically been utilized to distinguish the presence of CSF in bloody nasal discharge, especially after traumatic

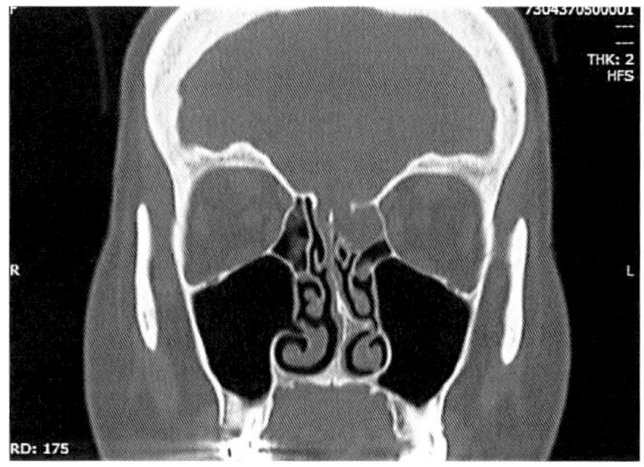

Fig. 3. High-resolution CT coronal reformat demonstrating a large left and smaller right cribriform defect in a woman with benign intracranial hypertension.

insult. A clear ring around a central stain of blood on a cloth is considered positive. However, tears and saliva may lead to a false-positive result [21]. Nasal endoscopy may be useful in localizing the site of defect or encephalocele during active CSF leakage demonstrating asymmetric glistening or herniation.

Differential Diagnosis

Ommaya [20] proposed a CSF rhinorrhea classification system describing traumatic and non-traumatic etiologies. Accidental trauma and surgical-related iatrogenic injury fall into the traumatic category with accidental trauma accounting for 70–80% of CSF rhinorrhea [22]. The nontraumatic category is further subdivided into high and normal intracranial pressures encompassing neoplasms, hydrocephalus, congenital defects, and erosive skull base processes. Roughly 45% of nontraumatic leaks are secondary to high-pressure causes, tumor being 84% of these cases [22]. Spontaneous or idiopathic CSF leak is a more recently defined category including patients with no traumatic history or underlying etiology [23, 24]. Young to middle-aged, obese females with benign intracranial hypertension and empty sella syndrome stereotype this cohort. The precise pathophysiology underlying the elevated intracranial pressure remains elusive, but the constant increased pulsatile pressure is thought to erode thin areas and the natural foramina of the skull base (fig. 3) [12].

Diagnostic Testing

Various methodologies have been utilized to discern the presence of CSF. Glucose content of a leak sample was used for decades with concentrations above 30 mg/dl considered diagnostic for CSF. The advent of glucose oxidase strips provided rapid assessment of glucose presence in CSF [25]. However, contaminants such as tears or nasal mucous decrease sensitivity and specificity [26]. Meurmann et al. [27] described beta-2-transferrin as a marker for CSF in 1979. Beta-2-transferrin is the di-sialated form of transferrin found only in CSF, perilymph, and ocular fluids, thereby providing a good marker for CSF [14]. Only 0.5 ml of fluid is needed, and reports demonstrate

97% sensitivity and 99% specificity [28, 29]. Beta-trace protein (B-TP) is another reliable marker of CSF, reported in 1961 by Clausen, which is synthesized by the choroid plexus and is the most abundant protein in CSF beside albumin and prealbumin [30]. Sensitivity of 91% and 100% specificity for CSF have been reported with this test [31]. Though B-TP may be slightly more reliable than beta-2-transferrin, serum protein concentrations of B-TP are significantly altered by bacterial meningitis and renal insufficiency [32]. Currently, beta-2-transferrin remains the assay of choice for confirming a sample of CSF rhinorrhea. This sample can be collected and stored under refrigeration for days while maintaining integrity for proper laboratory analysis [33].

Advancements in imaging have played a large role in the identification of CSF leaks. Dandy [18] first utilized plain films in 1926 to assess pneumocephalus in CSF leak patients [34]. Radionuclide cisternography involving intrathecal injection of a radionuclide tracer gained popularity in the 1970–1980s, but was limited as specific skull base defects could not be isolated. Endoscopically placed nasal pledgets absorbing the tracer agent over time with subsequent quantification of tracer activity improved sensitivity, reportedly up to 76% for active leaks [35]. Prolonged cisternography using indium-111 diethylene triaminepentaacetic acid allowed for intermittent CSF leaks assessment with 25% success rate in identifying intermittent leaks [36]. However, study data depends upon patient tolerance of intranasal pledgets, and even then, does not adequately localize the specific defect site [35]. Currently, radionucleotide cisternography is reserved for complex CSF leak cases when diagnosis is questionable despite other diagnostic tests [35, 36, 37].

High-resolution CT (HRCT) has become the primary imaging modality in the evaluation of a suspected CSF leak. As CT images with bone algorithm are able to visualize areas of osseous defect, active leaks do not need to be present for assessment. HRCT imaging has a reported sensitivity and specificity as high as 92 and 100%, respectively, and multiplanar reformats have identified defects as small as 2 mm [38, 39]. Sites of suspected CSF leak involve a skull base defect associated with mucosal thickening, soft tissue, or air-fluid level/opacification in the adjacent sinonasal space. Axial images are useful in assessing the posterior table of the frontal sinus and the posterolateral walls of the sphenoid sinus whereas coronal sections are beneficial for the cribriform plates and the sphenoid and ethmoid roofs (fig. 4) [34].

CT cisternography, developed in the 1970s, utilizes intrathecally contrast with CT imaging [40]. Pre- and postcontrast studies are obtained with the patient in trendelenberg position. Contrast in a sinus cavity or a 50% post-contrast Hounsfield unit increase is considered positive [35]. Sensitivity has been reported at 48%, increasing to 92% in active leaks [35, 38]. Lumbar puncture discomfort and complications such as headache, seizure or contrast-reaction limit this technique as a first-line radiographic study. Recent studies have also demonstrated that HRCT provides equivalent data with decreased morbidity [34, 35]. CT cisternography has been recommended after initial HRCT when multiple or broad-based skull-base defects are identified to assess which site is fistulous [34, 35].

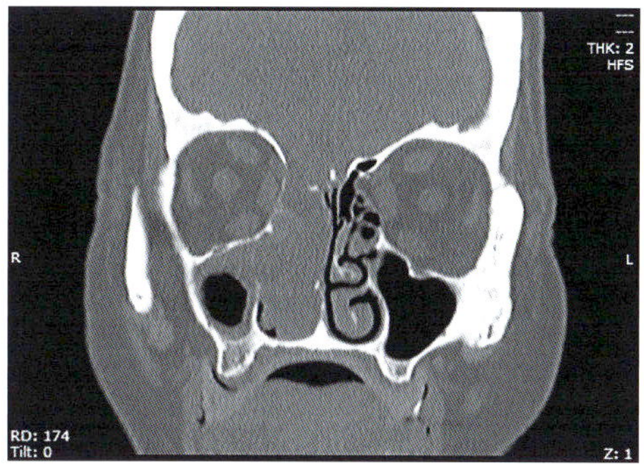

Fig. 4. High-resolution CT coronal reformat demonstrating a large encephalocele through a right ethmoid defect.

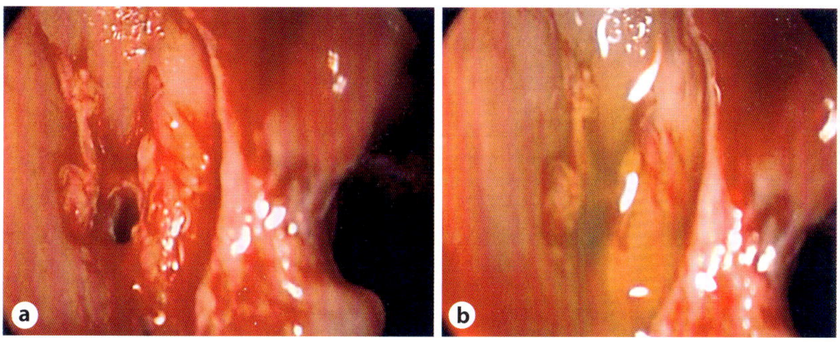

Fig. 5. a, **b** Intraoperative intrathecal fluorescein localizing site of skull base defect. **a** Skull base defect without intrathecal fluorescein. **b** Fluorescein shown leaking through defect.

MR imaging is useful when herniated intracranial contents or vasculature is a concern [34]. Studies have revealed sensitivities of 87% and 89% accuracy for MRI cisternography accounting for both active and inactive CSF leaks [38]. As with CT cisternography, MR cisternography is also useful for cases when multiple skull base defects are present. Though the advantage of meningocele or encephalocele identification is an important distinguishing feature for MR, HRCT is recommended as the initial imaging modality as it isolates the osseous defect [34].

Despite radiologic advances, the need to simultaneous detect and localize a CSF leak has led to the proliferation of intrathecal fluorescein testing (fig. 5). Kirchner and Proud [42] described the first intraoperative use of intrathecal fluorescein (IF) for CSF leak repair in the 1960s [41]. Sensitivities and specificities of 74 and 100%, respectively, have been reported. Additionally, IF has been beneficial in determining the adequacy of CSF leak repairs. However, use of IF is not FDA approved and is not without complication. Seizures, neurotoxicity, and allergic reactions among others have been reported [43, 44]. The frequency of these complications seems to be related to fluorescein concentration and speed of injection. A mixture of 0.1 ml of a 10% fluorescein solution with 9.9 ml of CSF injected slowly over 10 min has proven to be a safe method of use. Topically applied intranasal fluorescein has recently been described although its utility and ability to detect unexpected synchronous leaks remains to be seen [44, 45].

Conclusion

Skull base anatomy is a complex interface between the intracranial contents and the facial structures. Knowledge of the intimate relationships as well as the development of the skull base is integral for understanding the problems that manifest in this area. Defects of the skull base can lead to CSF leakage that may require operative intervention. A thorough history and clinical evaluation provides insight into potential etiology, location, and amount of CSF leak as well as complications. Multiple nonradiologic tests have been developed to confirm suspect rhinorrhea as CSF. Beta-2-transferrin remains the gold standard to date. Radiographic assessments have evolved to better define the precise location of CSF leaks, and HRCT remains the imaging technique of choice. For confirmation of CSF rhinorrhea and assurance of proper closure, intraoperative intrathecal fluorescein has become a recent valuable aid in endoscopic closure of skull base defects.

References

1 Keros P: Über die praktische Bedeutung der Niveauunterschiede der lamina cribrosa des ethmoids. Z Laryngol Rhinol Otol 1962;41:808–813.
2 Janecka IP, Tiedemann K (eds): Skull Base Surgery: Anatomy, Biology, and Technology. Lippincott-Raven, 1997.
3 Curtain HD, Som PM (eds): Head and Neck Imaging, ed 4. Philadelphia, Mosby, 2003.
4 Janfaza P, Nadol JB, Galla RJ, Fabian RL, Montgomery WM (eds): Surgical Anatomy of the Head and Neck. Cambridge, Harvard University Press, 2011.
5 Pinheiro-Neto CD, Fernandez-Miranda JC, Rivera-Serrano CM, Paluzzi A, Snyderman CH, Gardner PA, Sennes L: Endoscopic anatomy of the palatovaginal canal (palatosphenoidal canal): a landmark for dissection of the vidian nerve during endonasal transpterygoid approaches. Laryngoscope 2012;122:6–12.
6 Samii M, Draf W: Surgery of the Skull Base: an Interdisciplinary Approach. Berlin, Springer, 1989.
7 Friede H: Normal development and growth of the human neurocranium and cranial base. Scand J Plast Reconstr Surg 1981;15:163–169.
8 Hedlund G: Congenital frontonasal masses: developmental anatomy, malformations and MR imaging. Pediatr Radiol 2006;36:642–662.
9 Sternberg M: Ein bisher nicht beschriebener Kanal im Keilbein des Menschen. Anat Anz 1888;23:784–786.
10 Castelnuovo P, Dallan I, Pistochini A, Battaglia P, Locatelli D: Endonasal endoscopic repair of Sternberg's canal cerebrospinal fluid leaks. Laryngoscope 2007;117:345–349.
11 Bendersky DC, Landriel FA, Ajler PM, Hem SM, Carrizo AG: Sternberg's canal as a cause of encephalocele within the lateral recess of the sphenoid sinus: a report of two cases. Surg Neurol Int 2011;2:171.
12 Woodworth BA, Palmer JN: Spontaneous cerebrospinal fluid leaks. Curr Opin Otolaryngol Head Neck Surg 2009;17:59–65.
13 Davson H, Segal MB: Physiology of the CSF and Blood-Brain Barriers. Boca Raton, CRC Press, 1996.
14 Willis WD, Grossman RG: Medical Neurobiology, ed 3. Philadelphia, Mosby, 1981.
15 Malluci C, Sgouros S (eds): Cerebrospinal Fluid Disorders. News York, Informa Healthcare, 2010.
16 Zlab MK, Moore GF, Daly DT, Yonkers AJ: Cerebrospinal fluid rhinorrhea: a review of the literature. Ear Nose Throat J 1992;71:314–317.
17 Thomson St C: Cerebrospinal Fluid: Its Spontaneous Escape from the Nose. London, Cassell, 1899.
18 Dandy WE: Pneumocephalus, intracranial pneumocele or aerocele. Arch Surg 1926;12:949–982.

19 Brodie HA, Thompson TC: Management of complications from 820 temporal bone fractures. Am J Otol 1997;18:188–197.
20 Ommaya AK: Cerebrospinal fluid fistula and pneumocephalus; in Wilkins RH, Rengachary SS (eds): Neurosurgery. McGraw-Hill, 1996, pp 2773–2782.
21 Loew F: Traumatic, spontaneous and post-operative CSF rhinorrhea. Adv Tech Stand Neurosurg 1984; 11:169–207.
22 Kerr JT, Chu FWK, Bayles SW: Cerebrospinal fluid rhinorrhea: diagnosis and management. Otolaryngol Clin N Am, 2005;38:597–611.
23 Clark D, Bullock P, Hui T, Firth J: Benign intracranial hypertension: a cause of CSF rhinorrhea. J Neurol Neurosurg Psychiatry 1994;57:847–849.
24 Schlosser RJ, Woodworth BA, Wilensky EM, Grady MS, Bolger WE: Spontaneous cerebrospinal fluid leaks: a variant of benign intracranial hypertension. Ann Otol Rhinol Laryngol 2006;115:495–500.
25 Gadeholt H: The reaction of glucose-oxidase test paper in normal nasal secretion. Acta Otolaryngol (Stockh) 1964;58:271–272.
26 Katz RT, Kaplan PE: Glucose oxidase sticks and cerebrospinal fluid rhinorrhea. Arch Phys Med Rehabil 1985;66:391–393.
27 Meurmann OH, Irjala K, Suonpaa J, Laurent B: A new method for the identification of cerebrospinal fluid leakage. Acta Otoloaryngol 1979;87:366–369.
28 Skedros DG, Cass SP, Hisrch BE, KellyRH: Sources of error in use of beta-2-transferrin analysis for diagnosing perilymphatic and cerebral spinal fluid leaks. Otolaryngol Head Neck Surg 1993;109:861–864.
29 Warnecke A, Averbeck T, Wurster U, Harmening M, Lenarz T, Stover T: Diagnostic relevance of beta-2-transferrin for the detection of cerebrospinal fluid fistulas. Arch Otolaryngol Head Neck Surg 2004; 130:1178–84.
30 Raine C: Diagnosis and management of otologic cerebrospinal fluid leak. Otolaryngol Clin N Am 2005;38:583–895.
31 Bachmann G, Nekic M, Michel O: Clinical experience with beta-trace protein as a marker for cerebrospinal fluid. Ann Otol Rhinol Laryngol 2000; 109:1099–1102.
32 Meco C, Oberascher G, Arrer E, Moser G, Albegger K: Beta-trace protein test: new guidelines for the reliable diagnosis of cerebrospinal fluid fistula. Otolaryngol Head Neck Surg 2003;129:508–517.
33 Bleier BS, Debnath I, O'Connell BP, Vandergrift WA 3rd, Palmer JN, Schlosser RJ: Preliminary study on the stability of beta-2-transferrin in extracorporeal cerebrospinal fluid. Otolaryngol Head Neck Surg 2011;144:101–103.
34 Lloyd KM, DelGaudio JM, Hudgins PA: Imaging of skull base cerebrospinal fluid leaks in adults. Radiology 2008;248:725–736.
35 Stone JA, Castillo M, Neelon B, Mukherji SK: Evaluation of CSF leaks: high-resolution CT compared with contrast-enhanced CT and radionuclide cisternography. AJNR Am J Neuroradiol 1999;20: 706–712.
36 Glaubitt D, Haubrich J, Cordoni-Voutsas M: Detection and quantitation of intermittent CSF rhinorrhea during prolonged cisternography with ^{111}In-DTPA. AJNR Am J Neuroradiol 1983;4:560–563.
37 Lund VJ, Savy L, Lloyd G, Howard D: Optimum imaging and diagnosis of cerebrospinal fluid rhinorrhea. J Laryngol Otol 2000;114:988–992.
38 Shetty PG, Shroff MM, Sahani DV, Kirtane MV: Evaluation of high-resolution CT and MR cisternography in the diagnosis of cerebrospinal fluid fistula. AJNR Am J Neuroradiol 1998;19:633–639.
39 LaFata V, McLean N, Wise SK, DelGaudio JM, Hudgins PA: CSF leaks: correlation of high-resolution CT and multiplanar reformations with intraoperative endoscopic findings. AJNR Am J Neuroradiol 2008;29:536–541.
40 Drayer BP, Wilkins RH, Boehnke M, Horton JA, Rosenbaum AE: Cerebrospinal fluid rhinorrhea demonstrated by metrizamide CT cisternography. AJR Am J Roentgenol 1977;129:149–151.
41 Kirchner FR, Proud GO: Method for the identification and localization of cerebrospinal fluid, rhinorrhea and otorrhea. Laryngoscope 1960;70:921–931.
42 Seth R, Rajeskaran K, Benniger MS, Batra PS: The utility of intrathecal fluorescein in cerebrospinal fluid leak repair. Otolaryngol Head Neck Surg 2010; 143:626–632.
43 Schlosser RJ, Bolger WE: Nasal cerebrospinal fluid leak: critical review and surgical considerations. Laryngoscope 2004;114:255–265.
44 Ozturk K, Karabagli H, Bulut S, Egilmez M, Duran M: Is the use of topical fluorescein helpful for management of CSF leakage? Laryngoscope 2012;122: 1215–1218.
45 Jones ME, Reino T, Gnoy A, Guillory S, Wackym P, Lawson W: Identification of intranasal cerebrospinal fluid leaks by topical application with fluorescein dye. Am J Rhinol 2000;14:93–96.

Allen Lam
Department of Otolaryngology, Massachusetts Eye and Ear Infirmary
243 Charles Street
Boston, MA 02114 (USA)
Tel. +1 617 573 3654, E-Mail allen_lam@meei.harvard.edu

Bleier BS (ed): Comprehensive Techniques in CSF Leak Repair and Skull Base Reconstruction.
Adv Otorhinolaryngol. Basel, Karger, 2013, vol 74, pp 12–23 (DOI: 10.1159/000342264)

Pathophysiology of Iatrogenic and Traumatic Skull Base Injury

Stacey T. Gray · Arthur W. Wu

Department of Otology and Laryngology, Harvard Medical School, Massachusetts Eye and Ear Infirmary, Boston, Mass., USA

Abstract

Traumatic skull base injuries can be divided into iatrogenic injuries from surgery and accidental traumatic injuries. The most common cause of iatrogenic skull base injury is endoscopic sinus surgery. The reported rate of cerebrospinal fluid (CSF) leak as a complication of endoscopic sinus surgery is less than 1%. Intraoperative injury most commonly occurs at the ethmoid roof and the lateral lamella of the cribiform plate. Case complexity, such as revision surgery, as well as surgical experience has been shown to be a contributing factor to iatrogenic CSF leaks during endoscopic sinus surgery. Other causes of iatrogenic CSF leaks include endoscopic skull base and open neurosurgical procedures. CSF leak occurs in 2.6% of all closed head traumas and is more likely in the case of a definitive skull base fracture. The majority of accidental trauma CSF leaks can be managed conservatively, but those that persist beyond a week are typically managed surgically to avoid the risk of meningitis.

Cerebrospinal fluid leaks (CSF) can be divided into traumatic and non-traumatic causes. Traumatic etiologies can be further divided into surgical and accidental trauma. In this chapter, we will discuss the traumatic (iatrogenic and accidental) skull base injuries that together make up the majority of cases seen by otolaryngologists. Recognition of these injuries is imperative in order to avoid the potentially lethal complications of meningitis and pneumocephalus. The treatment of a skull base injury is dependent on the etiology, location, size and associated symptoms. Options for treatment can vary from observation to surgical exploration and closure. Both open and endoscopic procedures can be used for surgical closure of skull base defects. The experienced sinus surgeon has the advantage of a more direct and less invasive approach to closing these defects compared to the traditional neurosurgical approach via craniotomy, and endoscopic closure is often the initial option for management. Since the first endoscopic CSF leak repair was described by Wigand, many clinical series have confirmed this technique to have a >90% success rate [1–4]. However, neurosurgical approaches sometimes remain necessary, especially for larger and more diffuse injuries.

Traumatic/Iatrogenic Skull Base Injury

Iatrogenic CSF leaks are more common than accidental traumatic CSF leaks, making up 41–58% of CSF leaks in several series from large tertiary care centers [1–4]. Endoscopic sinus surgery, endoscopic skull base surgery, open skull base surgery, and other neurosurgical procedures can all result in inadvertent injury to the skull base with resultant CSF leak. Endoscopic sinus surgery (ESS) has become one of the most common otolaryngologic procedures in the United States, and ESS-related CSF leaks are more common than CSF leaks caused by skull base and neurosurgical procedures or by accidental trauma. ESS-related CSF leaks range from 25.9 to 41.7% of total CSF leaks [1, 3].

Endoscopic Sinus Surgery

While ESS is now a common surgical skill among most practicing otolaryngologists, it is still a relatively new technique. As Dr. Mosher is often quoted in regards to intranasal ethmoidectomy in the early 1920s, 'surgery in this region should be simple, but it has proven to be one of the easiest ways to kill a patient' [5]. The development of the endoscopic approach, improved understanding of sinus anatomy, and advances in technology and imaging studies have certainly improved the safety and success of sinus surgery, but a healthy respect for potential complication is necessary when planning surgery. Endoscopic sinus surgery is a safe procedure in the hands of well-trained surgeons, and many series have quoted a major complication rate of less than 1% [6]. However, in reviewing the literature, it is obvious that the reported complication rate should be carefully considered within the concept of a 'learning curve'. Dr. Stankiewicz's summary of his early experience with the endoscopic technique demonstrated an overall complication rate of 29% and a major complication rate of 8%; however, the study revealed a substantial individual improvement in the technique with increasing experience, reflected by a lower major complication rate of 2% in the latter half of the study [7]. Notably, the majority of the major complications were related to bleeding, and only one CSF leak was reported. In a recent follow-up study analyzing 25 years of practice and 3,402 patients, Stankiewicz et al. [8] found a reduced major complication rate of 3.1% and a CSF leak rate of 0.5%. It is important to remember that while the published complication rate for ESS in experienced hands is extremely low, a significant number of otolaryngologists have personally encountered a CSF leak during ESS. Platt et al. [9] distributed a questionnaire-based study to practicing otolaryngologists in New York and found that 25% of responders had experienced an intraoperative CSF leak in the 5 years prior to the study.

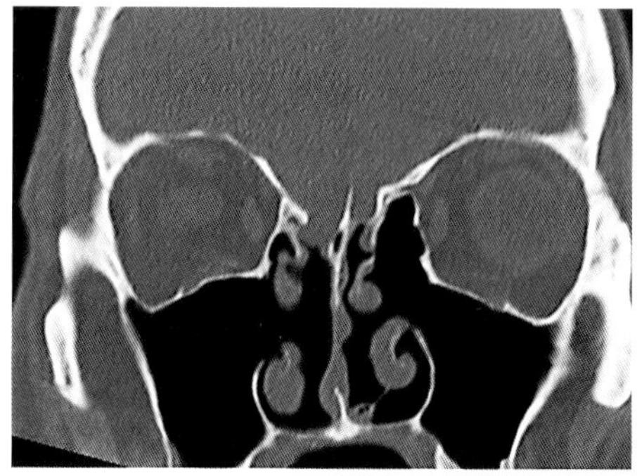

Fig. 1. Coronal CT scan demonstrating iatrogenic injury of the lateral lamella of the right cribiform plate after endoscopic sinus surgery.

Although iatrogenic CSF leak is uncommon during ESS, it is important to highlight some of the factors that might contribute to the development of this complication. Certain areas of the skull base are more prone to injury and understanding the complex anatomy of the sinuses is essential before proceeding with ESS. Review of pre-operative imaging studies is crucial, as these areas of potential concern can be scrutinized before surgery. In addition, it is important to remember that some cases might be more complex and more likely to result in injury to the skull base and should be approached with more caution. This includes cases where anatomy is significantly distorted, such as revision surgery or patients with significant polyposis, or in cases that require significant dissection at the level of the skull base.

The most common locations for iatrogenic injury to the skull base during ESS are the lateral lamella of the cribiform (fig. 1), the posterior fovea ethmoidalis, the frontal recess, and the sphenoid sinus. While determining the most frequent location for iatrogenic injury is partially academic, knowledge of the most dangerous areas of dissection is instructive to both the seasoned sinus surgeon and the surgeon in training. These areas of concern should be closely examined as part of the pre-operative assessment of imaging studies.

In a review of several large series, the ethmoid roof appears to be slightly more predominant in the frequency of injury during ESS compared to the cribiform plate (table 1) [1–3, 10, 11]. In order to understand why injury might occur at the ethmoid roof, it is important to review the skull base anatomy from a surgical perspective. As the skull base tends to slope inferiorly along the ethmoid roof from the frontal recess to the planum sphenoidale, it can be vulnerable to injury when working in an anterior to posterior direction. Using the technique first described by Dr. Messerklinger, entering the posterior ethmoid cavity inferiorly and moving posteriorly with caution is less likely to result in injury to the skull base. After locating the superior turbinate and sphenoid ostium, the skull base can easily be identified at the roof of the sphenoid sinus. Skeletonizing the skull base from a posterior to anterior direction can then be performed safely under direct visualization.

Table 1. Location of iatrogenic skull base Injuries during sinus surgery

Study	Ethmoid roof	Cribiform plate	Other
Tabaee et al. [1]	10 patients	8 patients	6 patients (sphenoid)
Zuckerman et al. [2]	14 patients	9 patients	
Banks et al. [3]	17 patients	9 patients	4 patients (frontal) 2 patients (sphenoid)
Smith et al. [16]	9 patients	8 patients	4 patients (sphenoid) 2 patients (frontal)
Lee et al. [11]	4 patients	0 patients	4 patients (frontal)

In only two series did the number of reported cribiform injuries outnumber the reported ethmoid roof injuries. In a review of the University of Pittsburgh experience with CSF leaks, the authors found that there were 19 cribiform injuries compared to 14 ethmoid roof injuries [4]. However, in this series, whether the traumatic injury was iatrogenic or caused by accidental trauma was not reported. In a recent review of his 25 years of experience, Dr. Stankiewicz comments that the majority of iatrogenic CSF leaks occurred at the lateral lamella of the cribiform plate, adjacent to the anterior ethmoid artery and just posterior to the frontal recess [8]. It is at this location that the skull base is thinnest. It is apparent why this area might be at risk. There is marked variation in the angle and depth by which the lateral lamella of the cribiform forms the superomedial border of the ethmoid sinuses. The Keros classification defines three classes of cribiform and skull base configurations with increasing depth, presuming that an increasing depth corresponds to increasing danger for potential injury: class I 1–3 mm, class II 3–7 mm, and class III 7–17 mm (fig. 2) [12]. It is also critical to assess whether the lateral lamella forms a gently or more steeply sloping angle. Additionally, the configuration of the cribiform may differ from side to side in a single patient, and the slope or depth of the skull base might be different on the contralateral side (fig. 2d, 3). This underscores the need for careful preoperative review of imaging studies to plan the surgical approach.

Iatrogenic skull base injuries at the sphenoid sinus or frontal recess are less common during ESS. Puncturing blindly into the sphenoid sinus may cause injury to the skull base in addition to the potentially devastating complications of carotid artery and optic nerve injury. The natural sphenoid ostium is the safest location for entry into the sphenoid sinus, as it is typically located below the level of the skull base [13]. The sphenoid sinus ostium is usually easily identifiable in the space between the superior turbinate and the septum. After the sphenoid ostium has been entered, the opening can be widened while the skull base is under direct vision in the interior of the sphenoid sinus. Pre-operative assessment for bony dehiscence within the sphenoid sinus, including the bone of the carotid artery and optic nerve is crucial to avoid potential injury. In addition, identification of anatomic variants, such as an Onodi cell

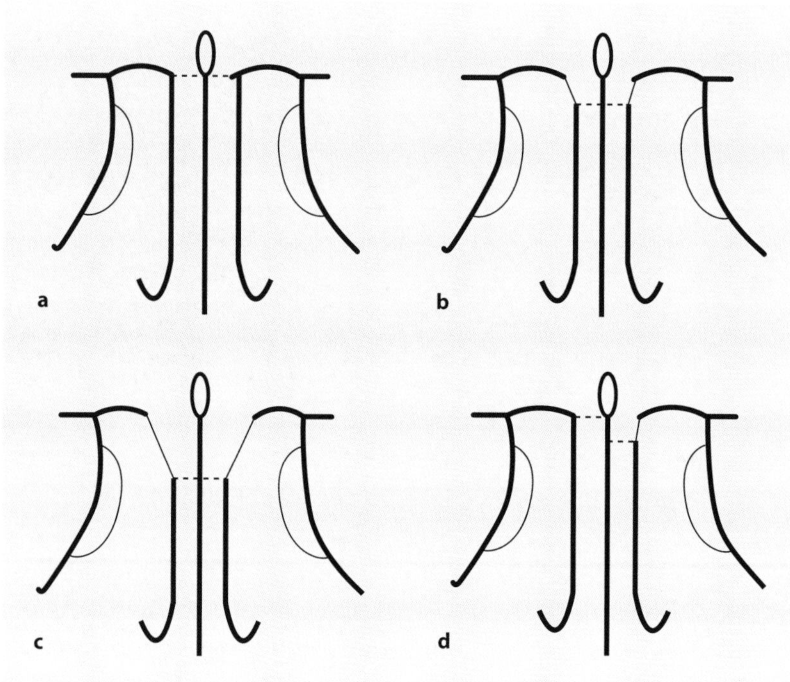

Fig. 2. Keros Classification of the Medial Skull Base Architecture. **a** Type I, 1–3 mm. **b** Type II, 3–7 mm. **c** Type III, 7–17 mm. **d** Asymmetric skull base.

(a posterior ethmoid cell that extends posterior to the sphenoid face), are important to recognize pre-operatively as this changes the configuration of the skull base. Most iatrogenic CSF leaks of the sphenoid sinus are due to transsphenoidal approaches to the sella. This will be covered in more depth in the section on neurosurgical procedures. Injuries to the frontal sinus during ESS are also very rare. Even in large series reviewing complicated frontal surgery such as drillout procedures, the rate of CSF leak is less than 1% [14, 15].

Identifying the typical surgical landmarks used during ESS is more difficult in revision surgery. Scarring and previous surgical resection of structures such as the middle turbinate may result in confusion for even the most experienced surgeon. This point is supported by the findings by Stankiewicz et al. [8] and Banks et al. [3] which showed that most CSF leaks occurred in revision ESS cases. In addition to careful pre-operative anatomic review, image guidance during surgery is often thought of as a way to potentially avoid complications during ESS. Smith et al. [16] reviewed the role of image guidance in avoiding complications during ESS. Currently, the literature on this topic consists of case series and expert opinions, so a definitive answer as to whether this technique improves complication rates is lacking. It is important to remember that image guidance should only be used as a confirmation tool of the

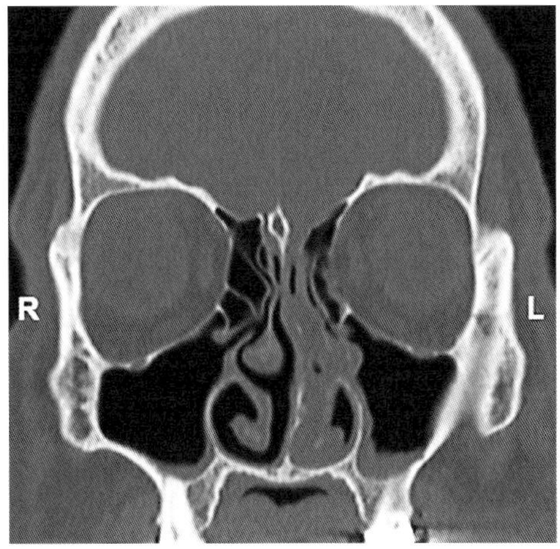

Fig. 3. Coronal CT demonstrating sloping ethmoid roof on left and flat ethmoid roof on right.

surgeon's own anatomical knowledge, rather than the deciding factor for identification of the skull base. Although not proven to decrease complication rates, most surgeons feel that image guidance is a helpful adjunct during revision ESS and can improve surgeon confidence [17].

Revision ESS in a patient that has a prior history of CSF leak repair is a case that is potentially ripe with complexity. In a series of revision ESS on patients with prior iatrogenic CSF leak, Reh et al. [18] found preoperative high-resolution CT and image guidance to be helpful in avoiding the previous area of repair. Patients with chronic rhinosinusitis with polyps may also be at higher risk for inadvertent skull base injury during ESS. These cases may be complicated by obscured anatomic landmarks, increased bleeding, and thinning or dehiscence of the skull base. Six of the 12 patients who underwent revision ESS with previous CSF leak in Reh et al.'s [18] case series had chronic rhinosinusitis with polyps, and 6 of the 16 CSF leaks in the review of Stankiewicz et al. [8] review of his career complications were also patients with polyps. In addition to the increased surgical complexity of ESS in patients with polyps, powered shavers are often used for dissection during these procedures. Church et al. [19] reported several cases of large skull base defects caused by powered shavers. Haphazard use of these instruments at the skull base is a potential cause of iatrogenic CSF leak.

Experience is also a contributing factor in the etiology of iatrogenic CSF leaks. Reasons that less experienced surgeons might be more likely to cause CSF leaks might seem obvious, such as lack of experience, but the less experienced also tend to be less surgically aggressive and may not attempt as thorough a dissection along the skull base. In a retrospective chart review of the experience with iatrogenic CSF leaks caused by ESS in a teaching hospital, Bumm et al. [20] found a correlation with the level of training and location of iatrogenic leaks. Junior otolaryngology residents were more prone to cause leaks at the cribiform, while senior residents and attending staff were more likely to cause leaks in the frontal recess. Surgeons of every experience level were equally likely to injure the posterior ethmoid roof. The overall rate of skull base injury with subsequent CSF leak was 0.46%. Junior residents had a CSF leak rate of 0.49%, but senior residents had a higher rate of 1.21%. At the senior resident level at the reporting institution, resident surgeons often operated for at least some portion of the case without the supervision of an attending, which could explain the increased risk at the senior resident level compared to the junior resident. Platt et al. [9] also found in a survey study that surgeons who performed more sinus surgeries per month had a statistically significant lower CSF leak rate than their peers who performed less than five surgeries per month. Limitations of this study were that the incidences were self-reported and based only on the recollection of the survey responder. Keerl et al. [21] also evaluated the association between experience and complications in the practice of several prominent sinus surgery pioneers. This reiterates the concept of the 'learning curve' as higher complication rates were noted earlier in these surgeon's careers and diminished with time. However, the most serious of complications occurred most frequently among experienced surgeons. In general, complication rates become less frequent with more experience, but serious complications still occur, even to the most seasoned surgeon.

Neurosurgical Procedures

A variety of neurosurgical procedures can result in CSF rhinorrhea if a dural defect is created at the time of surgery and not adequately repaired. This includes procedures involving the entire skull base from the posterior table of the frontal sinus to the clivus. Although surgery to address the actual skull base interface is the most obvious cause, other neurosurgical procedures, such as bifrontal craniotomy or aneurysm clipping can also result in an inadvertent CSF leak. CSF leak is a well-documented complication of neurosurgical procedures involving the skull base. In the past, it was somewhat uncommon for otolaryngologists to see these patients, as neurosurgeons often repaired these leaks through revision craniotomy. However, otolaryngology involvement in the closure of skull base defects has increased through the years as the techniques of endoscopic sinus surgery have advanced. In fact, the ability to assist with closure of CSF leaks likely contributed to the collaborative creation and advancement of the field of endoscopic skull base surgery over the last decade. In 2003, White et al. [22] reviewed a series of patients that underwent endoscopic closure of CSF leaks after neurosurgical procedures with an initial success rate of 92%. One patient had recurrent CSF leak after endoscopic repair and required revision endoscopic management, which was ultimately successful. The majority of the patients in this study experienced CSF leaks through the sphenoid sinus after pituitary surgery. However, the techniques can be applied to defects anywhere along the skull base.

Skull base procedures can be performed in different ways, depending on the site of the disease as well as the experience of the surgeon. Open approaches, including craniofacial resection and craniotomy are often employed, but endoscopic endonasal skull base surgery has gained popularity in recent years. Early experience with endoscopic endonasal surgery was primarily with transsphenoidal pituitary surgery, as this was similar to traditional transsphenoidal pituitary surgery performed with the microscope. Rates of CSF leak range from 2 to 13% in transseptal microscopic pituitary approaches, while purely endoscopic transsphenoidal techniques yield similar reported rates of 2–3.5% [23]. Open approaches to the anterior skull base via craniofacial resection (CFR) have a slightly higher rate of CSF leak. In a review of the Memorial Sloan-Kettering experience with open resections of malignant skull base lesions, Gil et al. [24] found an intracranial complication rate of 12–16%. Deschler et al. [25] reported a similar CSF leak rate of 13% after open anterior skull base surgery.

The CSF leak rate after endoscopic skull base surgery is widely variable in the literature. In the largest series from the University of Pittsburgh, the overall rate of postoperative CSF leak was 15.9%. Half of these procedures were approaches to sellar masses, such as pituitary adenomas, and the other half was more complicated cases [23]. However, there was a noted discrepancy between the CSF leak rate with early attempts at this technique and the CSF leak rate with more recent cases. This echoes the earlier discussion about the concept of the 'learning curve' of surgical experience. The authors attribute this improvement in postoperative CSF leak rate to the introduction of the vascularized, pedicled nasoseptal flap for closure. Prior to the use of this technique, the CSF leak rate was 58%, which was reduced to 5.4% after adoption of the nasoseptal flap. Other techniques for closure of endoscopically created skull base defects have also been used with success. In a study by Casiano's group, reviewing their experience of repairing large anterior skull base defects with alloderm, the CSF leak rate was only 3% [26]. Using the 'gasket seal' technique espoused by Anand and colleagues, large defects in the skull base have also been successfully repaired with low failure rates [27]. Tabaee et al. [28] presented a simple and logical algorithm for repair of surgically created skull base defects ranging from very small defects without CSF leak to large defects with intraoperative leaks. Very small defects without CSF leak were repaired with fascia, fat, or similar materials. Larger defects without violation of the arachnoid, such as pituitary adenoma resection, were reconstructed with gelfoam to fill the tumor cavity and septal vomer or cartilage to reconstruct the bony defect followed by tissue sealant over the reconstruction. Large defects with high flow CSF leaks were addressed with multilayered closures: a fat graft to fill the tumor cavity, a fascial underlay graft, a bone/cartilage graft to fill bony defect, and tissue sealant. Those defects in communication with the ventricular spaces did not use fat graft to fill the tumor cavity in order to avoid the theoretical complication of iatrogenic hydrocephalus [28]. A variety of options for closure exist, all with similar reported success, and ultimately the choice for repair is related to surgeon experience and preference. Additionally, the need for rigid reconstructive techniques of large defects in the era of vascularized flap repair remains controversial.

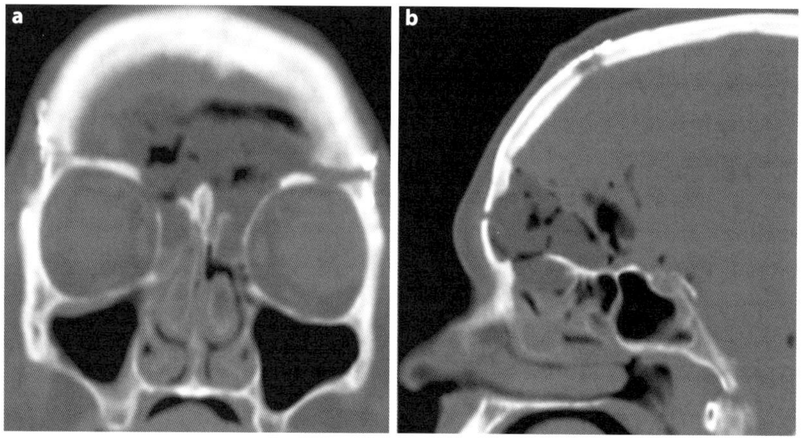

Fig. 4. Coronal (**a**) and sagittal (**b**) CT scan demonstrating pneumocephalus and CSF leak in patient who underwent frontal craniotomy, frontal sinus obliteration and pericranial flap reconstruction of a skull base defect after olfactory meningioma resection.

Other neurosurgical procedures also have potential for injury to the skull base, which might go unrecognized until the postoperative period. Accidental entry into the frontal sinus while performing bifrontal craniotomy can result in postoperative CSF rhinorrhea. In addition, inadequate obliteration of the frontal recess during frontal sinus cranialization can also result in a CSF leak postoperatively [29]. Even recognized skull base defects that are appropriately repaired at the time of surgery may fail and result in persistent CSF leaks. Figure 4 demonstrates the CT scan of a patient with a persistent CSF leak after open resection of an anterior skull base tumor and repair with a pericranial flap. In the previously mentioned review of endoscopic repair of CSF leaks caused by neurosurgical procedures, most CSF leaks were secondary to pituitary or hypophyseal surgery, but there was one case from frontal lobe debridement via frontal craniotomy and one case after circle of Willis aneurysm clipping [22]. In the exposure of clinoidal and supraclinoidal segments of the internal carotid artery, resection of the anterior clinoid process is necessary, and if there is pneumatization of the anterior clinoid process, the subarachnoid space can become contiguous with the sphenoid sinus [30]. Thus CSF rhinorrhea is a well-recognized complication of aneurysm surgery.

Traumatic/Accidental Skull Base Injury

The dura overlying the skull base is extremely adherent and trauma to this area can easily result in a tear in the dura. CSF leaks in the anterior skull base occur in 2.6% of all closed head trauma, but some minor leaks may be occult and resolve spontaneously

[29]. Overall, dural tears and CSF leaks occur in 10–30% of all skull base fractures. [28] The incidence of CSF leak after anterior skull base trauma is five to six times more common than in injuries to the posterior and lateral skull base [29, 31]. The danger of ascending meningitis from such trauma ranges from 20 to 57% [29, 31] and can result in relatively high mortality (10%). In a large review of their management of anterior skull base fractures, Scholsem et al. [29] found that the majority of patients (74%) had combined fractures involving multiple areas. As in other studies, the most common area involved in CSF fistula was the ethmoid-cribiform plate (81%) followed by the posterior table of the frontal sinus (77%), orbital roof (66%) and the sphenoid sinus (50%). Interestingly, the majority of patients (64%) in this study had mild head injuries as graded by the Glasgow Coma Scale (GCS) with scores from 13 to 15. Only 19% of patients had severe head injuries with GCS scores ≤8. Therefore, all closed head injuries even if mild in severity should be evaluated for CSF leak and injury to the skull base. The etiology of most injuries was secondary to motor vehicle accidents. The most common associated symptoms in this population were epistaxis, periorbital ecchymosis, chemosis, and rhinorrhea. The frontal sinus was almost always involved in the associated fractures, but other common associated injuries included zygomaticomaxillary complex fractures, Lefort II, and Lefort III midface fractures.

Frontal sinus fractures account for 5–15% of all maxillofacial fractures, and over half of the patients present with neurologic sequelae [32, 33]. Up to a third of all patients also present with CSF leaks. Usually the presence of a CSF leak reflects injury to the posterior table but can signal simultaneous fractures of the cribiform or the fovea ethmoidalis. Pneumocephalus may be seen on CT scan, but this is not always a sure sign of concurrent CSF leak. CT cisternography and endoscopic examination with intrathecal fluorescein are more specific methods of evaluation if there is concern for persistent CSF leak. 70–85% of CSF leaks associated with skull base fractures will spontaneously resolve with conservative management, but persistent leaks may need surgical intervention [32–34]. Prophylactic antibiotics are a controversial subject, but a meta-analysis by Brodie [35] demonstrated a significant improvement in the rate of meningitis in those with traumatic CSF leaks. Use of a lumbar drain may alleviate pressure over the injury and can be added after 48 h of conservative trial should the leakage persist [33, 34]. Those injuries that do not resolve within 1 week as well as injuries involving large defects should be managed surgically.

Conclusion

The majority of CSF rhinorrhea seen by the otolaryngologist is the result of inadvertent injury during endoscopic sinus surgery. The lateral lamella of the cribiform plate and the posterior fovea ethmoidalis are the areas that are most commonly injured. Less operative experience, revision surgery, and the presence of polyps add a potential increase risk for skull base injury during ESS. These issues can be potentially ameliorated with careful pre-operative planning using imaging studies, excellent knowledge of surgical anatomy, and persistent vigilance during surgery. Other iatrogenic etiologies of CSF rhinorrhea include various neurosurgical procedures, including endoscopic and open skull base surgery. Accidental traumatic skull base injuries to the anterior skull base are usually secondary to motor vehicle accidents and typically involve more than one anatomic site. Accidental traumatic skull base injuries often resolve with conservative management. Regardless of the etiology of the skull base injury, vigilance for the feared complications of ascending meningitis and pneumocephalus is always necessary. The management of iatrogenic and traumatic skull base injuries depends on the cause, the severity and the site of the injury, but can often be accomplished through an endoscopic approach.

References

1 Tabaee A, Kassenoff TL, Kacker A, Anand VK: The efficacy of computer assisted surgery in the endoscopic management of cerebrospinal fluid rhinorrhea. Otolaryngol Head Neck Surg 2005;133:936–943.
2 Zuckerman JD, DelGaudio JM: Utility of preoperative high-resolution CT and intraoperative image guidance in identification of cerebrospinal fluid leaks for endoscopic repair. Am J Rhinol 2008;22:151–154.
3 Banks CA, Palmer JN, Chiu AG, Kennedy DW, et al: Endoscopic closure of CSF rhinorrhea: 193 cases over 21 years. Otolaryngol Head Neck Surg 2009;140:826–833.
4 Zweig JL, Carrau RL, Kassam A, Hegazy H, et al: Endoscopic repair of cerebrospinal fluid leaks to the sinonasal tract: predictors of success. Otolaryngol Head Neck Surg 2000;123:195–201.
5 Mosher HP: The symposium of the ethmoid-a surgical anatomy of the ethmoidal labyrinth. Trans 34th Ann Meet Am Acad Ophthamology and Otolaryngology, 1929, vol 34, pp 376–410.
6 May M, Levine HL, Mester SJ, et al: Complications of endoscopic sinus surgery. Analysis of 2108 patients – incidence and prevention. Laryngoscope 1994;104:1080–1083.
7 Stankiewicz JA: Complications of endoscopic intranasal ethmoidectomy. Laryngoscope 1987;97:1270–1273.
8 Stankiewicz JA, Lal D, Connor M, Welch K: Complications in endoscopic sinus surgery for chronic rhinosinusitis: a 25-year experience. Laryngoscope 2011;121:2684–2701.
9 Platt MP, Shaye D, Parnes SM: Management of unexpected cerebrospinal fluid fistulae during endoscopic sinus surgery. Am J Rhinol 2007;21:611–614.
10 Lindstrom DR, Toohill RJ, Loehrl TA, Smith TL: Management of cerebrospinal fluid rhinorrhea: the Medical College of Wisconsin Experience. Laryngoscope 2004;114:969–974.
11 Lee DH, Lim SC, Joo YE: Treatment outcomes of endoscopic repairs of sinonasal cerebrospinal fluid leaks. J Craniofacial Surg 2011;22:1266–1270.
12 Keros P: On the practical value of differences in the level of the lamina cribosa of the ethmoid. Z Laryngol Rhinol Otol 1962;41:809–813.
13 Bolger WE, Keyes AS, Lanza DC: Use of the superior meatus and superior turbinate in the endoscopic approach to the sphenoid sinus. Otolaryngol Head Neck Surg 1999;120:308–313.
14 Samaha M, Cosenza MJ, Metson R: Endoscopic frontal sinus drillout in 100 patients. Arch Otolaryngol Head Neck Surg 2003;129:854–858.
15 Anderson P, Sindwani R: Safety and efficacy of the endoscopic modified Lothrop procedure: a systematic review and meta-analysis. Laryngoscope 2009;119:1828–1833.
16 Smith TL, Stewart MG, Orlandi RR, Setzen M, Lanza DC: Indications for image-guided surgery: the current evidence. Am J Rhinol 2007;21:80–83.
17 Kacker A, Tabaee A, Anand V: Computer-assisted surgical navigation in revision endoscopic sinus surgery. Otolaryngol Clin North Am 2005;38:473–482.
18 Reh DD, Metson R, Sindwani R: Sinus surgery in patients with previously repaired cerebrospinal fluid leaks. Arch Otolaryngol Head Neck Surg 2008;134:1187–1190.

19 Church CA, Chiu AG, Vaughan WC: Endoscopic repair of large skull base defects after powered sinus surgery. Otolaryngol Head Neck Surg 2003;129: 204–209.
20 Bumm K, Heupel J, Bozzato A, Iro H, Hornung J: Localization and infliction pattern of iatrogenic skull base defects following endoscopic sinus surgery at a teaching hospital. Ausis Nasus Larynx 2009;36:671–676.
21 Keerl R, Stankiewicz J, Weber R, Hosemann W, Draf W: Surgical experience and complications during endosnasal sinus surgery. Laryngoscope 1999;109: 546–550.
22 White DR, Dubin MG, Senior BA: Endoscopic repair of cerebrospinal fluid leaks after neurosurgical procedures. Am J Otolaryngol 2003;24:213–216.
23 Kassam AB, Prevedello DM, Carrau RL, Snyderman CH, et al: Endoscopic endosnasal skull base surgery: analysis of complications in the authors' initial 800 patients. J Neurosurg 2011;114:1544–1568.
24 Gil Z, Patel SG, Bilsky M, Shah JP, Kraus DH: Complications after craniofacial resection for malignant tumors: are complication trends changing? Otolaryngol Head Neck Surg 2009;140: 218–223.
25 Deschler DG, Gutin PH, Mamelak AN, McDermott MW, Kaplan MJ: Complications of anterior skull base surgery. Skull Base Surg 1996;6:113–118.
26 Germani RM, Vivero R, Herzallah IR, Casiano RR: Endoscopic reconstruction of large anterior skull base defects using acellular dermal allograft. Am J Rhinol 2007;21:615–618.
27 Leng LZ, Brown S, Anand VK, Schwartz TH: 'Gasket-seal' watertight closure in minimal-access endoscopic cranial base surgery. Neurosurgery 2008;62(5 suppl 2):ONSE342–343.
28 Tabaee A, Anand VK, Brown SM, Lin JW, Schwartz TH: Algorithm for reconstruction after endoscopic pituitary and skull base surgery. Laryngoscope 2007;117:1133–1137.
29 Scholsem M, Scholtes F, Collignon F, Martin D, et al: Surgical management of anterior cranial base fractures with cerebrospinal fluid fistulae: a single-institution experience. Neurosurgery 2008; 62:463–471.
30 Chi JH, Sughrue M, Kunwar S, Lawton MT: The 'Yo-Yo' technique to prevent cerebrospinal fluid rhinorrhea after anterior clinoidectomy. Neurosurgery 2006;59(ONS suppl 1):ONS101–107.
31 Bernal-Sprekelsen M, Bleda-Vazquez C, Carrau RL: Ascending meningitis secondary to traumatic cerebrospinal fluid leaks. Am J Rhinol 2000;14:257–259.
32 Manolidis S, Hollier LH: Management of frontal sinus fractures. Plast Reconstr Surg 2007;120:32S–48S.
33 Ommaya AK: Cerebrospinal fluid fistula and pneumocephalus; in Wilkins RH, Rengachary SS (eds): Neurosurgery. New York, McGraw-Hill, 1996, pp 2773–2782.
34 Bell RB, Dierks EJ, Homer L, Potter BE: Management of cerebrospinal fluid leak associated with craniomaxillofacial trauma. J Oral Maxillofac Surg 2004; 62:676–684.
35 Brodie HA: Prophylactic antibiotics for posttraumatic cerebrospinal fluid fistulae. A meta-analysis. Arch Otolaryngol Head Neck Surg 1997;123:749–752.

Stacey T. Gray
Department of Otology and Laryngology, Harvard Medical School, Massachusetts Eye and Ear Infirmary
243 Charles Street
Boston, MA 02114 (USA)
E-Mail stacey_gray@meei.harvard.edu

Bleier BS (ed): Comprehensive Techniques in CSF Leak Repair and Skull Base Reconstruction.
Adv Otorhinolaryngol. Basel, Karger, 2013, vol 74, pp 24–32 (DOI: 10.1159/000342265)

Grafting Materials in Skull Base Reconstruction

Kara K. Prickett · Sarah K. Wise

Emory University Department of Otolaryngology, Atlanta, Ga., USA

Abstract

Successful endoscopic repair of the skull base may be performed with a variety of grafting materials. Graft materials discussed in this chapter are broadly categorized as autologous tissue, acellular human dermis, engineered collagen products, rigid support materials and tissue glues. Autologous tissues continue to be widely used due to their safety, availability, and low cost. Engineered collagen products and acellular dermis are favored in revision cases and larger repairs. The need for rigid support continues to be an area of controversy as pedicled grafts are increasingly used for larger defects. Tissue glues and sealants, however, are used by many authors. The components of, and evidence for, commonly used sealants are detailed. Relatively little high-quality, comparative evidence exists to guide decision making when selecting graft materials, but where available, this evidence is cited. If surgeon experience and patient characteristics are carefully considered, excellent results can be expected with most materials.

As endoscopic management has evolved to be considered standard of care for many skull base defects, successful repair has been described with multiple techniques and materials. Studies have shown that endogenous tissues, allografts, engineered xenografts, and synthetic materials may all be used successfully for closure of cerebrospinal fluid (CSF) leaks. Many authors describe multilayered closures using several different materials, which may include tissue glues and sealants or additional rigid support. Relatively little comparative data is available to guide decision making when choosing among available graft materials. Therefore, patient factors, defect characteristics, cost, and functional goals must be considered in the development of individual reconstruction plans. This chapter will review the classes of commonly used graft materials and, when available, present evidence to help guide choice of graft material in endoscopic skull base reconstruction.

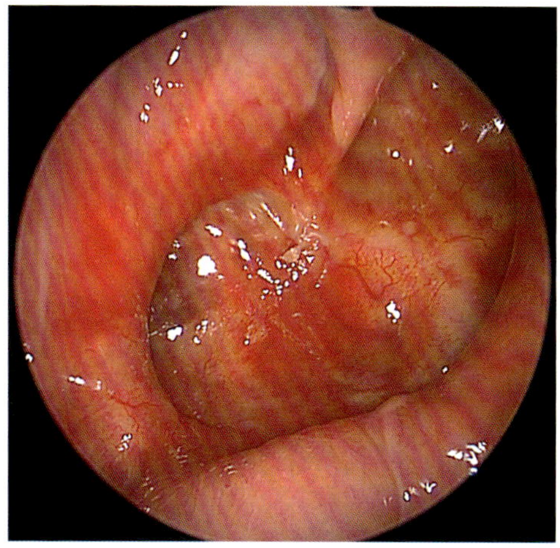

Fig. 1. Endoscopic view of the clival region 26 months following chordoma resection. Closure of this skull base defect was performed with abdominal fat and an inferior turbinate mucosal free graft. Likely resulting from postoperative radiation therapy, this patient has mild chronic mucosal inflammation in the region of mucosal repair.

Autologous Tissues

Successful skull base repairs have been reported using fat, fascia, bone, cartilage and free mucosal grafts. Fat is typically used in a 'plug' fashion, either to fill dead space or to obliterate the sphenoid sinus. Fascia may be placed as an underlay between the dura and skull base, or as an overlay in the sinus cavities. Multilayered fascia lata grafts have proven especially effective in management of open cisterns and other high-flow CSF leaks [1]. Bone or cartilage grafts are typically used in an underlay fashion to support larger repairs, and will be further discussed in the section on rigid support.

Because the current literature consists of retrospective series that do not control for use of adjuvant materials (fat, bone, tissue glues, overlay with pedicled mucosal graft) or methods (use of lumbar drain, type of absorbable and nonabsorbable packing), comparison of success rates with different autologous materials is not possible. However, articles published within the last 15 years report closure rates greater than 90% on first attempt, with an increase to 97–100% on second attempt [2–4].

Autologous tissues are popular due to their availability, biocompatibility and negligible cost; only the donor site morbidity and increase in OR time needed for harvest must be considered. Free mucosal grafts, fascia, fat, and bone or cartilage can all be harvested with minimal difficulty or advanced planning. Autologous grafts also have the benefit of rapid incorporation into the surrounding tissues (fig. 1). In a 1999 study of mucosal wound healing, Hoseman et al. [5] found that free middle turbinate mucosa grafts were fixed to the skull base after 6 days, and microscopic evidence of re-epithelialization was seen by postoperative day 12. For patients with processes that require follow-up with serial imaging, the absence of chronic inflammation at the repair site can help avoid ambiguity and unnecessary biopsies.

Disadvantages of autologous grafts include limited tissue supply and the need to create a separate incision or wound. While major complications are infrequent, the donor site may be associated with pain, minor complications, and additional healing time. Free mucosal grafts have also been noted to contract by nearly 20% during the healing process [5]. This can lead to repair failure if the edges of the graft are not integrated into the surrounding mucosa when the contraction occurs.

Acellular Human Dermis

Acellular human dermis (Alloderm®, LifeCell Corporation, Branchburg, N.J., USA) has been used in many areas of head and neck surgery, and has gained popularity for reconstruction of anterior and middle cranial fossa defects over the past decade. Multilayered closure with acellular human dermis after transsellar surgery results in successful closure of CSF leak in 90–100% of cases on first attempt [6, 7]. No operative time is used for graft harvest, and the reconstitution process for the material requires only 10 minutes of soaking. Advantages of human dermis include its low propensity for shrinkage and its ease of use. Crusting can be an issue during the healing period, and debridement may be required for upwards of 8 weeks [8]. There is a theoretical risk of transmission of prion-mediated diseases whenever allogeneic tissue implants are used, but the skin is not known to be a reservoir for infectious material in Creutzfeldt-Jacob disease (CJD) [9]. More than 2 million patients have received acellular dermis grafts without a known case of transmission of CJD [10].

Engineered Collagen Products

Advances in tissue engineering have given rise to a proliferation of collagen-based dural replacement products. These products, often termed collagen 'sponges', 'fleeces', or 'matrices', consist of collagen-rich animal tissues that have been processed to remove all cells and antigens. Many products are then chemically treated to induce cross-linking among the collagen fibrils, thereby creating porous 'matrix' structures. Exact treatment processes and resultant product characteristics vary by manufacturer, but in general, the microstructure of each product is optimized to encourage fibroblast in-growth, angiogenesis, and new collagen formation [11]. Table 1 lists trade names and manufacturers of collagen matrix products that have been studied in skull base reconstruction. Matrix products are often used alone for reconstruction of smaller defects, or may be used in conjunction with pedicled or free mucosal overlays for closure of larger defects.

The products are designed to fully incorporate into native tissues via collagen remodeling while minimizing the surrounding inflammatory response. A recent study by the authors of this chapter examined short-term healing after skull base repair, showing more rapid mucosalization and resolution of crusting with engineered collagen products than with acellular human dermis [8]. Histopathologic studies of

Table 1. Collagen matrix products with published reports of use in endoscopic skull base repair

Product	Manufacturer	Components
Duraform®*	Codman & Schurtleff, Inc., Raynham, Mass., USA	type I collagen from bovine Achilles tendon
DuraGen®*	Integra Neurosciences, Plainsboro, N.J., USA	type I collagen from bovine Achilles tendon; ground, then chemically cross-linked
Duraguard®*	Synovis, St. Paul, Minn., USA	glutaraldehyde cross-linked bovine pericardium
Duramatrix®*	Stryker, Kalamazoo, Mich., USA	type I collagen from bovine Achilles tendon
Durasis®*	Cook Medical, Bloomington, Ind., USA	non-cross-linked collagen biomatrix from porcine small intestine
Durepair®*	Medtronic, Minneapolis, Minn., USA	type I and type III collagen matrix from acellular fetal bovine skin
Lyoplant®*	Aesculap, Inc., Center Valley, Pa., USA	lyophilized type I collagen matrix from bovine pericardium
TachoComb®+	Takeda Pharmaceuticals (recently purchased Nycomed), Deerfield, Ill., USA	equine collagen mesh coated with human fibrinogen and thrombin
TissuDura®+	Baxter Biomedical, Vienna, Austria	colloidal type I collagen from minced equine Achilles tendon

*Registered with the Food and Drug Administration.
+Not currently available in the United States.

TissuDura® (Baxter Biomedical, Vienna, Austria), an equine Achilles tendon-based product, showed inflammation and fibroblast infiltration by 4 weeks, collagen formation by 8 weeks, and full incorporation into surrounding dura by 16 weeks [12–14] (fig. 2). Zerris et al. [15] compared the tensile strength and tissue incorporation characteristics of Duraguard® (Synovis, St. Paul, Minn., USA), DuraGen® (Integra Neurosciences, Plainsboro, N.J., USA), and Durepair® (Medtronic, Minneapolis, Minn., USA) in a canine model. They found Duraguard and Durepair to be as strong as native dura, while DuraGen was quite fragile. This likely represents differences in tissue processing, and DuraGen is indeed recommended as an onlay graft, while suturing is recommended for Durepair and Duraguard. The looser structure of DuraGen (pore size 100 μm) was associated with faster replacement by host tissue (incorporated by 4 weeks) than either Durepair (pore size 10–20 μm, populated with host cells by 6 months) or Duraguard (pore size 5 μm, no host fibroblast population by 6 months) [15].

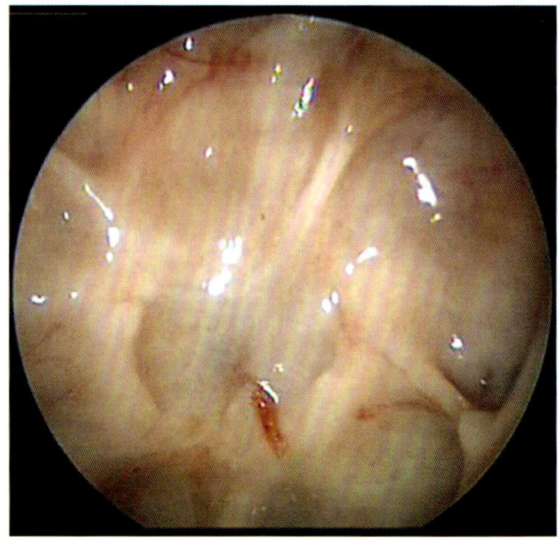

Fig. 2. Endoscopic view of the sella 8 months' status postendoscopic transsphenoidal pituitary adenomectomy and reconstruction of the bony sella defect with engineered collagen matrix.

Though some degree of inflammatory response must take place for collagen matrix products to be resorbed, severe inflammation is rare. In a study of TissuDura® (Baxter Biomedical, Vienna, Austria), Biroli et al. [16] noted severe inflammation on post-op imaging in only 1 of 114 patients. The lack of associated inflammation is critical for repairs within or adjacent to the central nervous system, as inflammatory responses may not only lead to ambiguity in postoperative imaging, but may also theoretically act as epileptogenic foci.

Rigid Support

Indications for use of rigid support in reconstruction of the skull base are controversial. Some authors feel that all CSF leak repairs should have rigid support, while others favor use of vascularized flaps to prevent the formation of sequestra that can be associated with large rigid grafts [17, 18]. Strict size criteria for defects that require rigid support have not been agreed upon: Nyqyist et al. [19] advise use of a 'gasket seal' type closure with a soft tissue underlay held in place by bone for defects greater than 1 cm^2, while Tabaee et al. [20] base their repairs on which layers of the skull base have been breached (bone, arachnoid, ventricle, etc.) rather than on the size of the defect alone. Generally, authors agree that rigid support should be used to buttress repairs in patients with elevated intracranial pressure whenever possible, as this population has a higher incidence of repair failure [21–23].

Autologous and synthetic materials are available for rigid support. Vomer and septal cartilage are frequently employed as they can be harvested with minimal donor site morbidity, but use of conchal cartilage, split calvarial grafts, rib, and iliac crest have been described. Synthetic materials can be grouped into 'bone substitutes', such as polymethylmethacrylate or hydroxyapatite and 'plates', such as those made from titanium mesh or expanded polytetrafluoroethylene. Methylmethacrylate products are somewhat easier to use than hydroxyapatite products due to their faster hardening times (approximately 10 vs. 30 min) and their ability to harden in wet fields [24, 25]. Published reports note that both materials are associated with minimal inflammatory response and a low risk of extrusion [24, 25]. Titanium mesh is malleable, easily seen on imaging, and also quite safe. Esposito et al. [17] found migration or dislodgement of the mesh to be problematic in only 3 of 378 patients with a first-attempt CSF leak closure rate of 97.5%. This closure rate corresponds to rates published by other groups who do not routinely use rigid support, so the question of whether or not to repair a defect with rigid material remains one that must be answered on a case-by-case basis.

Tissue Glues

The vast majority of endoscopic skull base repair techniques described in the current literature include the use of tissue glues or sealants. Commonly used materials are shown in table 2 [26–28]. Use of these materials is idiosyncratic and typically off-label; safety and efficacy trials involving brain and spine surgery have been conducted, but limited information pertaining to endoscopic surgery is available. Chin et al. [29] published a small series using a polyethylene glycol-based hydrogel (DuraSeal® Covidien, Mansfield, Mass., USA) in endoscopic skull base surgery, and had successful CSF leak repair in 4 of 5 patients on first attempt with no complications. de Almeida et al. [30] attempted to quantify the benefit of adding fibrin glue (Tisseal®, Baxter Healthcare, Deerfield, Ill., USA) to a typical open repair using a pericranial flap and found that the addition of fibrin glue resulted in a higher likelihood of graft adherence at 7 days after surgery and an increase in 9 psi in burst pressure of the repairs. Conversely, with the advent of vascularized overlay mucosal reconstruction, the benefits of tissue glues are less clear. Eloy et al. [31] reported a series of 74 high-flow CSF leak repairs using a vascularized septal flap with (n = 42) or without (n = 32) a dural sealant. The only postoperative failure occurred in the sealant group, suggesting that the use of tissue glues may not provide an added benefit. Manufacturer inserts for many sealant products contain warnings about their use in enclosed spaces, as the materials tend to swell after reconstitution [27]. The potential for extravasation into the intracranial compartment should be carefully considered before using any sealant product during endoscopic skull base surgery.

Choice of Material

Given the current state of evidence, a stepwise approach to skull base repair, such as that advocated by Tabaee et al. [20], seems most appropriate. Simple defects may be closed in simple fashion. As multilayered techniques have not been shown to be

Table 2. Commonly used dural sealants [26, 27]

Product	Manufacturer	Components	Current data
DuraSeal®	Covidien, Waltham, Mass., USA	– polyethylene glycol-trilysin amine	– FDA approved spine sealant – safe in human and animal studies – no difference in CSF leak rate vs. controls
BioGlue®	Cryolife, Kennesaw, Ga., USA	– purified bovine serum albumin – glutaraldehyde	– not approved for use in neurosurgical procedures – used in 32 sellar re-constructions without complication* – noted to be neurotoxic by manufacturer
Evicel®	Johnson & Johnson Wound Management Ethicon, Inc., Sommerville, N.J., USA	– biological active component 2 (cryoprecipitate) – human thrombin	– not approved for use in neurosurgical procedures – subdural application safe in animal studies – no large human studies available in neurosurgery or otolaryngology
Tisseal®	Baxter Healthcare Corporation, Deerfield, Ill., USA	– sealer protein concentrate (cryoprecipitate) – human thrombin	– not approved for use in neurosurgical procedures – several large studies of off-label use have failed to show neurologic complications – direct application to vessels may cause venous occlusion

*Study funded by Cryolife [28].

superior to single-layer closures, more complex techniques should be reserved for complex defects that involve several structures [32]. Cost and increased morbidity should be carefully considered to ensure they are commensurate with patient benefit.

An ideal material for repair of skull base defects would be widely available, easily handled, inexpensive, carry no risk of systemic toxicity, and associated with minimal local inflammation or immune response [16]. It would also serve equally well as a dural substitute and a scaffold for remucosalization of the sinonasal cavity. While this product may not yet be on the market, careful technique and conscientious selection of available materials yield excellent results in most patients. Several large, retrospective studies have failed to show an association between material choice and outcome of skull base repair [32, 33]. Prospective, comparative evidence is lacking, however. Well-designed, multicenter studies of materials used in endoscopic skull base repair could greatly benefit this rapidly expanding field.

References

1 Luginbuhl AJ, Campbell PG, Evans J, Rosen M: Endoscopic repair of high-flow cranial base defects using a bilayer button. Laryngoscope 2010;120: 876–880.

2 Gilat H, Rappaport Z, Yaniv E: Endoscopic transnasal cerebrospinal fluid leak repair: a 10 year experience. Isr Med Assoc J 2011;13:597–600.

3 Liu P, Wu S, Li Z, Wang B: Surgical strategy for cerebrospinal fluid rhinorrhea repair. Neurosurgery 2010;66:281–285.

4 Sciarretta V, Mazzatenta D, Ciarpaglini R, Pasquini E, Farneti G, Frank G: Surgical repair of persisting CSF leaks following standard or extended endoscopic transsphenoidal surgery for pituitary tumor. Minim Invas Neurosurg 2010;53:55–59.

5 Hosemann W, Goede U, Sauer M: Wound healing of mucosal autografts for frontal cerebrospinal fluid leaks – clinical and experimental investigations. Rhinology 1999;37:108–112.

6 Citardi MJ, Cox AJ 3rd, Bucholz RD: Acellular dermal allograft for sellar reconstruction after transsphenoidal hypophysectomy. Am J Rhinol 2000;14: 69–73.

7 Ismail AS, Costantino PD, Sen C: Transnasal transsphenoidal endoscopic repair of CSF leakage using multilayer acellular dermis. Skull Base 2007;17: 125–132.

8 Prickett KK, Wise SK, Delgaudio JM: Choice of graft material and postoperative healing in endoscopic repair of cerebrospinal fluid leak. Arch Otolaryngol Head Neck Surg 2011;137:457–461.

9 WHO Guidelines on Tissue Infectivity Distribution in Transmissible Spongiform Encephalopathies. 2006.

10 Warren WL, Medary MB, Dureza CD, et al: Dural repair using acellular human dermis: experience with 200 cases: technique assessment. Neurosurgery 2000;46:1391–1396.

11 Cappabianca P, Esposito F, Cavallo LM, et al: Use of equine collagen foil as dura mater substitute in endoscopic endonasal transsphenoidal surgery. Surg Neurol 2006;65:144–148.

12 Gazzeri R, Neroni M, Alfieri A, et al: Transparent equine collagen biomatrix as dural repair. A prospective clinical study. Acta Neurochir (Wien) 2009;151:537–543.

13 Knopp U, Christmann F, Reusche E, Sepehrnia A: A new collagen biomatrix of equine origin versus a cadaveric dura graft for the repair of dural defects – a comparative animal experimental study. Acta Neurochir (Wien) 2005;147:877–887.

14 Narotam PK, van Dellen JR, Bhoola KD: A clinicopathological study of collagen sponge as a dural graft in neurosurgery. J Neurosurg 1995;82:406–412.

15 Zerris VA, James KS, Roberts JB, Bell E, Heilman CB: Repair of the dura mater with processed collagen devices. J Biomed Mater Res B Appl Biomater 2007;83:580–588.

16 Biroli F, Fusco M, Bani GG, et al: Novel equine collagen-only dural substitute. Neurosurgery 2008; 62:273–274.

17 Esposito F, Dusick JR, Fatemi N, Kelly DF: Graded repair of cranial base defects and cerebrospinal fluid leaks in transsphenoidal surgery. Neurosurgery 2007;60:295–303.

18 Harvey RJ, Nogueira JF, Schlosser RJ, Patel SJ, Vellutini E, Stamm AC: Closure of large skull base defects after endoscopic transnasal craniotomy. J Neurosurg 2009;111:371–379.

19 Nyquist GG, Anand VK, Mehra S, Kacker A, Schwartz TH: Endoscopic endonasal repair of anterior skull base non-traumatic cerebrospinal fluid leaks, meningoceles, and encephaloceles. J Neurosurg 2010;113:961–966.

20 Tabaee A, Anand VK, Brown SM, Lin JW, Schwartz TH: Algorithm for reconstruction after endoscopic pituitary and skull base surgery. Laryngoscope 2007;117:1133–1137.

21 Wang EW, Vandergrift WA 3rd, Schlosser RJ: Spontaneous CSF Leaks. Otolaryngol Clin North Am 2011;44:845–856.

22 Woodworth BA, Prince A, Chiu AG, et al: Spontaneous CSF leaks: a paradigm for definitive repair and management of intracranial hypertension. Otolaryngol Head Neck Surg 2008;138: 715–720.

23 Wise SK, Schlosser RJ: Evaluation of spontaneous nasal cerebrospinal fluid leaks. Curr Opin Otolaryngol Head Neck Surg 2007;15:28–34.

24 Costantino PD, Hiltzik DH, Sen C, et al: Sphenoethmoid cerebrospinal fluid leak repair with hydroxyapatite cement. Arch Otolaryngol Head Neck Surg 2001;127:588–593.

25 Moliterno JA, Mubita LL, Huang C, Boockvar JA: High-viscosity polymethylmethacrylate cement for endoscopic anterior cranial base reconstruction. J Neurosurg 2010;113:1100–1105.

26 Summary of safety and effectiveness data: DuraSeal. 2004.

27 Epstein NE: Dural repair with four spinal sealants: focused review of the manufacturers' inserts and the current literature. Spine J 2010;10:1065–1068.

28 Kumar A, Maartens NF, Kaye AH: Reconstruction of the sellar floor using Bioglue following transsphenoidal procedures. J Clin Neurosci 2003;10:92–95.
29 Chin CJ, Kus L, Rotenberg BW: Use of duraseal in repair of cerebrospinal fluid leaks. J Otolaryngol Head Neck Surg 2010;39:594–599.
30 de Almeida JR, Ghotme K, Leong I, Drake J, James AL, Witterick IJ: A new porcine skull base model: fibrin glue improves strength of cerebrospinal fluid leak repairs. Otolaryngol Head Neck Surg 2009;141:184–189.
31 Eloy JA, Choudhry OJ, Friedel ME, Kuperan AB, Liu JK: Endoscopic nasoseptal flap repair of skull base defects: is addition of a dural sealant necessary? Otolaryngol Head Neck Surg 2012;147:161–166.
32 Banks CA, Palmer JN, Chiu AG, O'Malley BW Jr, Woodworth BA, Kennedy DW: Endoscopic closure of CSF rhinorrhea: 193 cases over 21 years. Otolaryngol Head Neck Surg 2009;140:826–833.
33 Zweig JL, Carrau RL, Celin SE, et al: Endoscopic repair of cerebrospinal fluid leaks to the sinonasal tract: predictors of success. Otolaryngol Head Neck Surg 2000;123:195–201.

Sarah K. Wise, MD
Emory Sinus Nasal and Allergy Center, Department of Otolaryngology
550 Peachtree St., MOT 9th floor
Atlanta, GA 30308 (USA)
E-Mail sarah.wise@emoryhealthcare.org

Bleier BS (ed): Comprehensive Techniques in CSF Leak Repair and Skull Base Reconstruction.
Adv Otorhinolaryngol. Basel, Karger, 2013, vol 74, pp 33–41 (DOI: 10.1159/000342266)

Free Graft Techniques in Skull Base Reconstruction

Jonathan Y. Ting · Ralph Metson

Department of Otolaryngology, Massachusetts Eye and Ear Infirmary and Department of Otology and Laryngology, Harvard Medical School, Boston, Mass., USA

Abstract

Over the last 20–30 years, the management of anterior skull base defects, CSF leaks and encephaloceles has shifted towards endoscopic, minimally invasive approaches. Whether their etiology is spontaneous or traumatic, the large majority of these lesions can be repaired with free mucosal grafts. Such grafts may be readily harvested from the nasal septum or turbinates and applied as either a single layer reconstruction for small leaks or as a multilayer reconstruction for larger defects. The accessibility, ease of placement, and high take rate associated with free mucosal grafts makes them a wise and popular choice for reconstruction of many skull base defects.

With the widespread adoption of endoscopic sinonasal surgical techniques and instrumentation over the last 20–30 years, the management of anterior skull base defects, cerebrospinal fluid (CSF) leaks and nasal meningoencephaloceles has shifted away from open craniotomy incisions and towards endoscopic, minimally invasive approaches. Since Wigand's report in 1981 of employing a transnasal endoscopic technique to repair a CSF leak, multiple reports have described the refinement and high success rate of this method [1]. Endoscopic approaches offer the advantage of enhanced visualization and magnification, thereby facilitating defect identification and graft placement. By avoiding the need for open craniotomy and frontal lobe retraction, transnasal approaches also reduce the risk of anosmia, seizures, memory deficits, and intracranial hemorrhage associated with neurosurgical repairs from above. No matter which endoscopic grafting methods are utilized, endoscopic repair of CSF leaks have proved effective with success rates greater than 90% in most studies [2–6].

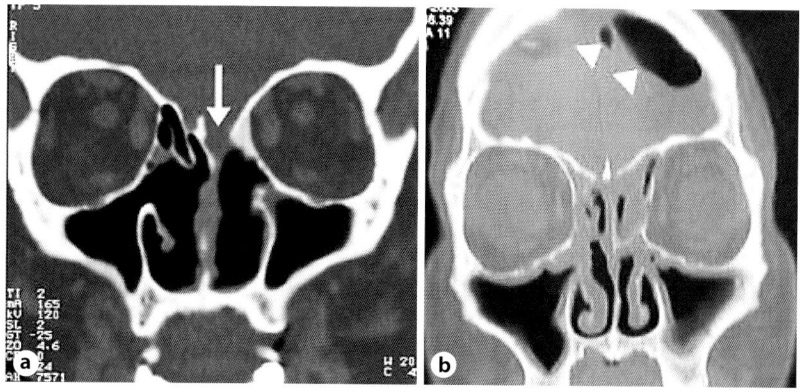

Fig. 1. **a** CT scan of a patient with a CSF leak following sinus surgery reveals a defect the cribriform plate (arrow). **b** CT scan of a patient with mental status changes following endoscopic sinus surgery demonstrated pneumocephalus (arrowheads). A skull base defect was identified in the anterior ethmoid roof. Endoscopic repair with a free mucosal graft harvested from the nasal septum resulted in resolution of the CSF leak, resorption of the intracranial air, and return of normal neurologic function.

The etiology of CSF leaks and skull base defects can be classified into traumatic and nontraumatic causes. Traumatic etiologies include accidental or iatrogenic trauma from either a planned skull base defect after tumor resection or complications during endoscopic sinus surgery. Inadvertent penetration of the skull base leads to injury of the underlying dura with CSF leak in approximately 0.5% of the surgical cases [7]. The bone of the skull base is thinnest and most susceptible to injury at the junction of the anterior ethmoid artery and the middle turbinate along the anterior ethmoid roof [8]. However, injury may occur anywhere along the skull base, including the posterior ethmoid roof, cribriform plate, posterior wall of the frontal sinus, and roof of the sphenoid sinus (fig. 1). Iatrogenic CSF leaks should be addressed at the time of surgery and surgeons performing sinus surgery should be comfortable performing a primary repair of this complication.

Nontraumatic forms of CSF leak include congenital defects, sinonasal tumors or other erosive processes, as well as spontaneous/idiopathic cases. Spontaneous leaks are more likely in patients with benign intracranial hypertension and often occur along the anterior cribiform plate or lateral recess of the sphenoid sinus (fig. 2) [9]. The elevated CSF pressures in this group of patients result in high rates of encephalocele formation and higher rates of recurrence [10–12]. Successful repair may require utilization of lumbar drains and acetazolamide perioperatively as well as the placement of underlay bone grafts. Patients often present with clear rhinorrhea (fig. 3), and diagnosis can be confirmed with beta-2-transferrin testing of a fluid sample.

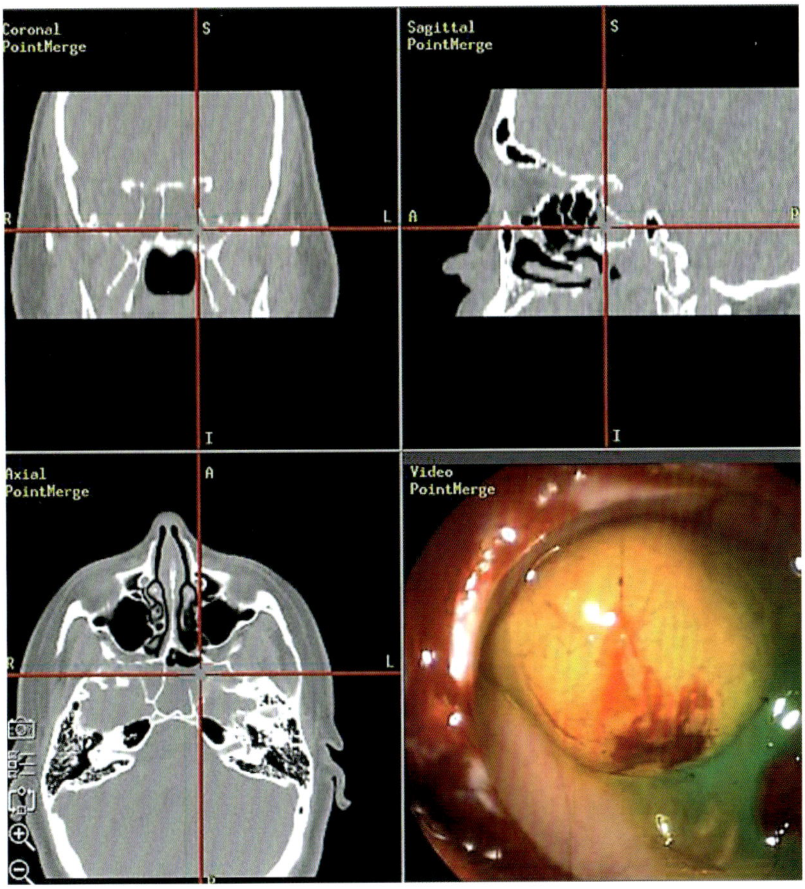

Fig. 2. Triplanar CT scan views and intraoperative endoscopic view of a lateral sphenoid recess encephalocele with preoperative instillation of flourescein into the intrathecal space. This skull base defect was repaired with a two-layer closure – a septal cartilage underlay graft and a free mucosal overlay graft.

Selection of Graft Material

Advantages of free grafts over pedicled grafts for repair of skull base defects include limited harvest time, relative technical ease of harvest, customized sizing, lack of tethering or risk of retraction by a vascular pedicle, ease of placement in any location that can be visualized and accessed with endoscopic instruments, and high long-term success rates. Potential benefits of vascularized flaps over free grafts included a generally thicker flap and better healing in irradiated or soon-to-be irradiated wound beds.

Despite lacking an axial vessel, free mucosal grafts have been shown to heal primarily with surrounding mucosa even when placed directly on maxillary bone [13]. Such grafts may be used in isolation for closure of small (<10 mm) defects [14] or in

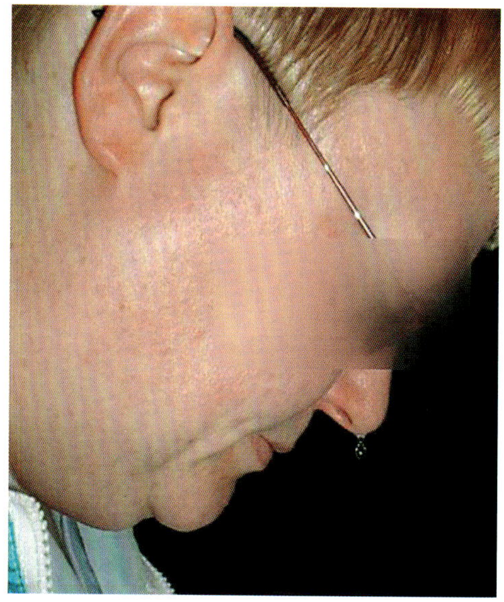

Fig. 3. Clear rhinorrhea upon leaning forward characteristic of a CSF leak.

combination with a larger multilayered closure [15]. Endoscopic repair of CSF leaks with free mucosal grafts result in high success rates of closure (93–95%) [4].

A large variety of free graft materials have been described, including temporalis fascia, fascia lata, muscle plugs, abdominal fat, cartilage grafts (from the septum or auricle), bone grafts (from the nasal septum or calvarium), as well as acellular dermal allografts and collagen dural substitutes. Interestingly, the material used for grafting does not seem to influence the likelihood of success [3, 6]. In a meta-analysis of 289 CSF fistulas, Hegazy et al. [3] found that the choice of the material did not alter the outcome of the repair. Selection of grafting material thus largely depends on the experience and preference of the surgeon.

Intraoperative Technique

When a CSF leak occurs during surgery, it often appears as a stream of darker fluid coursing through red blood. If only minimal bleeding is present, alternating streams of clear CSF and red blood are seen. When confronted with an unexpected intraoperative CSF leak, it is best for the surgeon to pack the nose, step back, and collect his or her thoughts. When recognized at the time of surgery and managed correctly, such complications have a greater than 90% likelihood of success with good clinical outcome [3].

In cases of delayed recognition of CSF leak or in cases of blunt trauma or encephalocele repair, a lumbar puncture with injection of intrathecal fluorescein may facilitate identification of the defect. A total of 0.1 ml of preservative-free 10% fluorescein is mixed with 10 ml of CSF and slowly injected into the intrathecal space. It is important to discuss and obtain written consent from the patient preoperatively for the use of fluorescein: specifically, mention should be made that fluorescein is not FDA-approved for intrathecal injection, because seizures and neurotoxicity have been reported at higher concentrations and with more rapid injections.

While defects along the ethmoid roof and the midline parasellar sphenoid sinus are readily accessible with standard endoscopic techniques, certain locations such as the lateral sphenoid and frontal sinuses may be more difficult to address. Additionally, CSF leaks along the olfactory cleft may involve multiple foramina and the placement of an underlay graft along the thin, oblique bone of the cribiform plate may be technically challenging. Certain locations in the lateral sphenoid or superolateral frontal sinus require extended endoscopic procedures. Lateral recess sphenoid defects involving the middle cranial fossa most often require an endoscopic transpterygoid approach with removal of the posterior maxillary wall and dissection of the pterygomaxillary fossa [16]. Posterior frontal table defects require at least wide a frontal sinusotomy for visualization. A decision must be made regarding the ability of repairing the leak endoscopically while ensuring long-term patency of the frontal outflow tract. Frontal drillout/modified endoscopic Lothrop procedures may assist with maintaining patency after repair as well as providing improved exposure. However, defects that are superior and lateral to the reach of frontal sinus instrumentation require either trephine or osteoplastic flap for access [17].

If a meningoencephalocele is present, bipolar cautery may be used to fulgurate the intranasal portion until its entrance through the skull base is identified. If a lumbar drain has been placed, the drain is opened to decrease ICP and aid in graft placement. Several millimeters of bone surrounding the defect should be exposed to facilitate mucosal grafting.

Free mucosal grafts can be harvested from the septum [18, 19], turbinates [20] as well as the nasal floor. Septal mucosa harvest results in exposed cartilage that may take longer to heal, increases the risk of septal perforation, and could potentially compromise the future use of a nasoseptal flap [21]. Harvesting mucosa from an intact middle turbinate is technically challenging and may increase the risk of causing a CSF leak at the superior attachment of the middle turbinate [22]. If the middle turbinate is resected for exposure, harvest of the middle turbinate mucosal graft on the back table has no additional morbidity. Resection of the middle turbinate does however increase the risk of bleeding from branches of the sphenopalatine. Similarly, a branch of the inferior turbinate artery runs along the medial aspect of the inferior turbinate in 60% of cadaveric specimens [23], but mucosa harvested from the inferior turbinate does seem to be thicker than mucosa from the middle turbinate. Regardless of mucosal harvest site, it is important to harvest a larger piece than anticipated to account for contraction of the graft after harvest. Healing of the defect site may take several weeks to months, with complete remucosalization typically achieved within 90 days [24].

Precise and meticulous graft placement is essential for successful endoscopic repair. A multilayer closure consisting of underlay grafts of fascia and cartilage, as well as an overlay graft of mucosa has been described (fig. 4) [15]. The underlay graft of cartilage, fascia, acellular dermal allograft, or collagen dural substitute is placed on the intracranial side of the defect, and is held in place by the ambient intracranial pressure. Mucosal grafts should never be placed intracranially as intracranial mucocele after endoscopic CSF leak repair has been reported, likely from improper mucosal graft placement [25]. Inking the mucosal surface of the graft with a surgical marker may facilitate correct mucosa placement, with the glandular mucus-secreting side directed toward the nasal cavity (fig. 5).

For most intraoperative CSF leaks, the senior author has found that repair with a single layer consisting of a free intranasal mucosal onlay graft harvested from the septum or turbinate is all that is necessary to seal the defect. Occasionally, larger defects require an additional structural layer, such as septal bone or cartilage placed on the intracranial side of the defect. The mucosal graft is stabilized and protected with a generous layer of absorbable packing material such as Surgicel (Johnson & Johnson, Pistacaway, N.J., USA) or Gelfoam (Pfizer, New York, N.Y., USA). Surgical sealants such as fibrin glue may also be used to reinforce the defect. An additional layer of nonabsorbable packing, such as gauze strip or Merocel (Medtronic, Jacksonville, Fla., USA) may be placed to further secure the repair and provide hemostasis. The nonabsorbable packing is removed after 1 week, whereas the absorbable packing is left in place and allowed to dissolve (a video demonstrating this technique is available online: see online suppl. video 1).

OSM

Other techniques include the 'bath-plug' technique described by Wormald and McDonogh [26] that incorporates an intracranial fat graft held in place by the ambient ICP, as well as the 'gasket seal' closure described by Leng et al. [27]. This closure consists of a soft tissue graft such as fascia lata centered over the defect with the edges of the graft exceeding the bony defect circumferentially. Next, a rigid graft such as vomer bone or Medpor (Porex Surgical, Newnan, Ga., USA) roughly the size of the bony defect is fashioned and gently countersunk into the bony defect, forming a watertight 'gasket seal'.

Postoperative Management

Once the patient is awake and out of the operating room, a CT scan may be obtained to assess for pneumocephalus or intracranial injury. Patients are typically observed overnight. Nonabsorbable nasal packing is removed at the 1 week postoperative visit, and patients are kept on antistaphylococcal antibiotics until packing is removed. The site of surgical repair is generally not debrided. Patients are seen again at 3–4 weeks to ensure adequate healing. Activity is limited to light exercise for 8 weeks and patients are prescribed stool softeners postoperatively to prevent straining. A CT scan may be performed 3–6 months after surgery for surveillance of the repair.

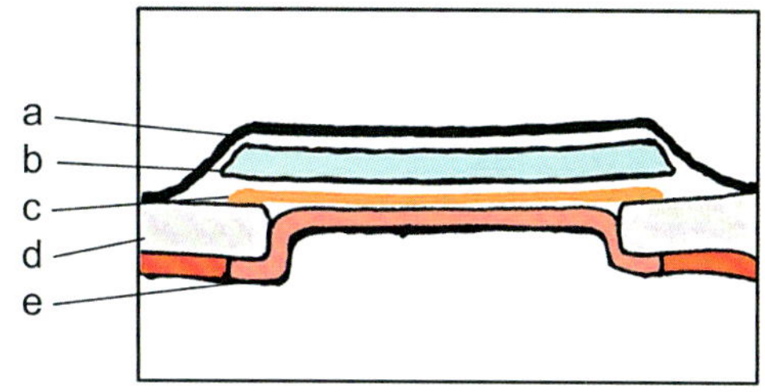

Fig. 4. Schematic of multilayer closure of skull base defect using free grafts: dura (**a**); cartilage or bone graft (**b**); fascia or alloderm graft (**c**); surrounding skull base (**d**); free mucosal graft (**e**).

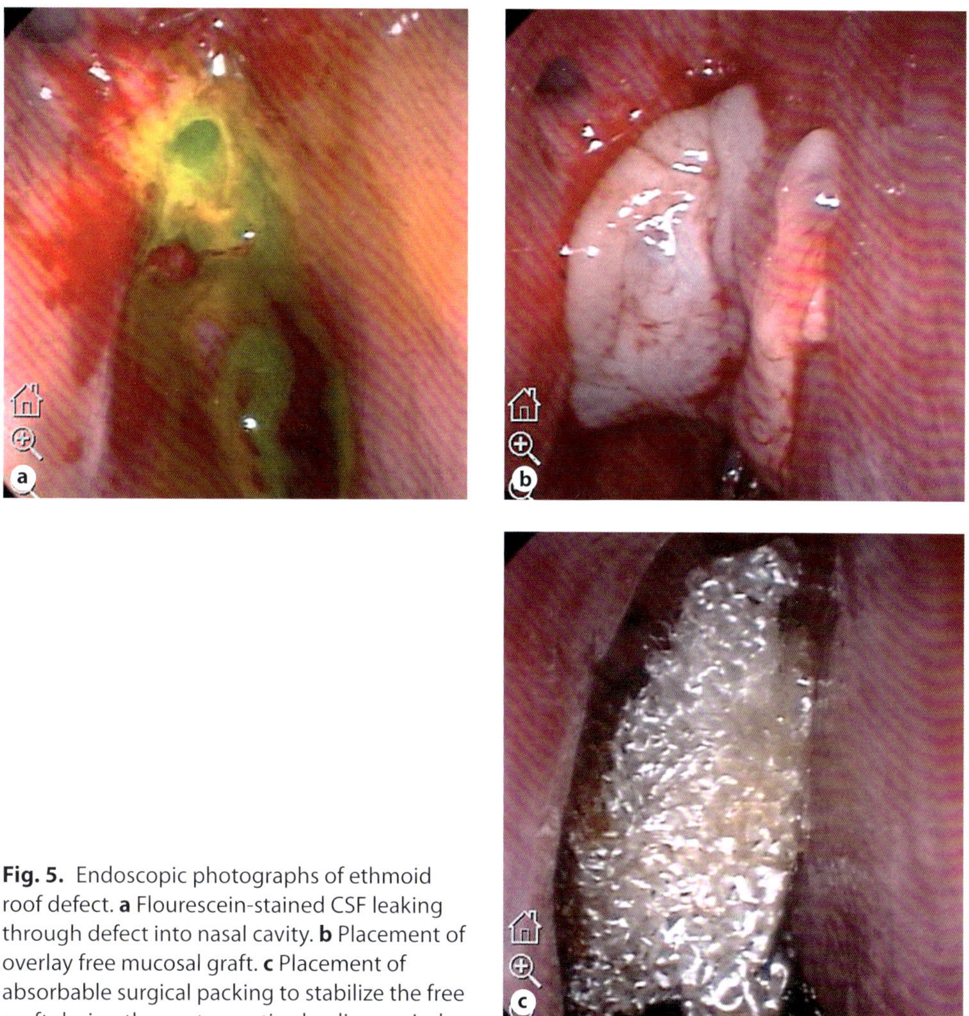

Fig. 5. Endoscopic photographs of ethmoid roof defect. **a** Flourescein-stained CSF leaking through defect into nasal cavity. **b** Placement of overlay free mucosal graft. **c** Placement of absorbable surgical packing to stabilize the free graft during the postoperative healing period.

Conclusion

Over the last 20–30 years, the management of anterior skull base defects, CSF leaks and nasal encephaloceles has shifted towards endoscopic, minimally invasive procedures, with success rates exceeding 90%. The majority of small anterior skull base defects can be repaired with free mucosal grafts, although large defects in irradiated beds may benefit from pedicled flap reconstruction.

References

1 Wigand ME: Transnasal ethmoidectomy under endoscopical control. Rhinology 1981;19:7–15.

2 Banks CA, Palmer JN, Chiu AG, O'Malley BW Jr, Woodworth BA, Kennedy DW: Endoscopic closure of CSF rhinorrhea: 193 cases over 21 years. Otolaryngol Head Neck Surg 2009;140:826–833.

3 Hegazy HM, Carrau RL, Snyderman CH, Kassam A, Zweig J: Transnasal endoscopic repair of cerebrospinal fluid rhinorrhea: a meta-analysis. Laryngoscope 2000;110:1166–1172.

4 Lanza DC, O'Brien DA, Kennedy DW: Endoscopic repair of cerebrospinal fluid fistulae and encephaloceles. Laryngoscope 1996;106:1119–1125.

5 Senior BA, Jafri K, Benninger M: Safety and efficacy of endoscopic repair of CSF leaks and encephaloceles: a survey of the members of the American Rhinologic Society. Am J Rhinol 2001;15:21–25.

6 Zweig JL, Carrau RL, Celin SE, et al: Endoscopic repair of cerebrospinal fluid leaks to the sinonasal tract: predictors of success. Otolaryngol Head Neck Surg 2000;123:195–201.

7 Bhatti MT, Giannoni CM, Raynor E, Monshizadeh R, Levine LM: Ocular motility complications after endoscopic sinus surgery with powered cutting instruments. Otolaryngol Head Neck Surg 2001; 125: 501–509.

8 Kainz J, Stammberger H: The roof of the anterior ethmoid: a place of least resistance in the skull base. Am J Rhinol 1989;3:191–199.

9 Schlosser RJ, Bolger WE: Significance of empty sella in cerebrospinal fluid leaks. Otolaryngol Head Neck Surg 2003;128:32–38.

10 Gassner HG, Ponikau JU, Sherris DA, Kern EB: CSF rhinorrhea: 95 consecutive surgical cases with long term follow-up at the Mayo Clinic. Am J Rhinol 1999;13:439–447.

11 Hubbard JL, McDonald TJ, Pearson BW, Laws ER Jr: Spontaneous cerebrospinal fluid rhinorrhea: evolving concepts in diagnosis and surgical management based on the Mayo Clinic experience from 1970 through 1981. Neurosurgery 1985;16:314–321.

12 Schick B, Ibing R, Brors D, Draf W: Long-term study of endonasal duraplasty and review of the literature. Ann Otol Rhinol Laryngol 2001;110: 142–147.

13 Mahendran S, Stevens-King A, Yung MW: How we do it: the viability of free mucosal grafts on exposed bone in lacrimal surgery – a prospective study. Clin Otolaryngol 2006;31:324–327.

14 Casiano RR, Jassir D: Endoscopic cerebrospinal fluid rhinorrhea repair: is a lumbar drain necessary? Otolaryngol Head Neck Surg 1999;121:745–750.

15 Lorenz RR, Dean RL, Hurley DB, Chuang J, Citardi MJ: Endoscopic reconstruction of anterior and middle cranial fossa defects using acellular dermal allograft. Laryngoscope 2003;113:496–501.

16 Lai SY, Kennedy DW, Bolger WE: Sphenoid encephaloceles: disease management and identification of lesions within the lateral recess of the sphenoid sinus. Laryngoscope 2002;112:1800–1805.

17 Woodworth BA, Schlosser RJ, Palmer JN: Endoscopic repair of frontal sinus cerebrospinal fluid leaks. J Laryngol Otol 2005;119:709–713.

18 Mattox DE, Kennedy DW: Endoscopic management of cerebrospinal fluid leaks and cephaloceles. Laryngoscope 1990;100:857–862.

19 Schlosser RJ, Bolger WE: Nasal cerebrospinal fluid leaks: critical review and surgical considerations. Laryngoscope 2004;114:255–265.

20 Lee TJ, Huang CC, Chuang CC, Huang SF: Transnasal endoscopic repair of cerebrospinal fluid rhinorrhea and skull base defect: ten-year experience. Laryngoscope 2004;114:1475–1481.

21 Suh JD, Ramakrishnan VR, DeConde AS: Nasal floor free mucosal graft for skull base reconstruction and cerebrospinal fluid leak repair. Ann Otol Rhinol Laryngol 2012;121:91–95.

22 Prevedello DM, Barges-Coll J, Fernandez-Miranda JC, et al: Middle turbinate flap for skull base reconstruction: cadaveric feasibility study. Laryngoscope 2009;119:2094–2098.

23 Lee HY, Kim HU, Kim SS, et al: Surgical anatomy of the sphenopalatine artery in lateral nasal wall. Laryngoscope 2002;112:1813–1818.
24 de Almeida JR, Snyderman CH, Gardner PA, Carrau RL, Vescan AD: Nasal morbidity following endoscopic skull base surgery: A prospective cohort study. Head Neck 2011;33:547–551.
25 Wang L, Kim J, Heilman CB: Intracranial mucocele as a complication of endoscopic repair of cerebrospinal fluid rhinorrhea: case report. Neurosurgery 1999;45:1243–1245.
26 Wormald PJ, McDonogh M: 'Bath-plug' technique for the endoscopic management of cerebrospinal fluid leaks. J Laryngol Otol 1997;111:1042–1046.
27 Leng LZ, Brown S, Anand VK, Schwartz TH: 'Gasket-seal' watertight closure in minimal-access endoscopic cranial base surgery. Neurosurgery 2008;62:ONSE342–343.

Jonathan Y. Ting
Department of Otolaryngology, Massachusetts Eye and Ear Infirmary and
Department of Otology and Laryngology, Harvard Medical School
243 Charles Street,
Boston MA 02114 (USA)
E-Mail jonathan_ting@meei.harvard.edu

OSM

Bleier BS (ed): Comprehensive Techniques in CSF Leak Repair and Skull Base Reconstruction.
Adv Otorhinolaryngol. Basel, Karger, 2013, vol 74, pp 42–55 (DOI: 10.1159/000342271)

Nasoseptal Flap

Carlos D. Pinheiro-Neto[a] · Carl H. Snyderman[a,b]

[a]Department of Otolaryngology, [b]Department of Neurological Surgery, University of Pittsburgh School of Medicine, Pittsburgh, Pa., USA

Abstract

The nasoseptal flap had an important role in the development of endoscopic endonasal surgery of the cranial base. The flap is pedicled upon the posterior septal artery, which is a terminal branch of the sphenopalatine artery. The reliable vascular supply promotes rapid healing and the flap is an effective barrier for the prevention of CSF leaks. For large skull base defects, it has dramatically decreased the risk of a postoperative CSF leak to less than 5%. The nasoseptal flap is a versatile flap with a wide arc of rotation that allows the flap to reach defects from the frontal sinus to the lower clivus. This chapter provides a comprehensive review of the nasoseptal flap. The harvesting and reconstruction techniques, modifications of the flap and complications are described.

The evolution of endoscopic endonasal surgery of the cranial base is intrinsically related to advancements in reconstruction. Just as the pericranial flap has been for open cranial base surgery, the cornerstone of endonasal reconstruction is the nasoseptal flap. The use of vascularized flaps to separate the nasal cavity from the intracranial space minimizes the risk of postoperative cerebrospinal fluid (CSF) leaks, pneumocephalus, and meningitis, especially in patients with prior radiation therapy or anticipated need for postoperative radiation therapy.

Prior to adaptation of the nasoseptal flap for endonasal skull base reconstruction, large dural defects of the anterior cranial base were repaired with multiple layers of non-vascularized tissues: fascia (autologous and cadaveric), fat, cartilage or bone, and alloplastic materials. These were supplemented with suturing techniques, nasal packing or balloon catheters for external support, and perioperative spinal drains to lower CSF pressure. With high flow intraoperative leaks, the incidence of postoperative CSF leaks was too high (20–30%) and a valid deterrent to the propagation of endonasal skull base surgery. With the advent of the vascularized mucosal nasoseptal flap, the postoperative CSF leak rate dramatically decreased to less than 5% [1–3].

In the last decade, alternatives to the nasoseptal flap have been described, consisting of vascularized local and regional flaps. Local flaps are harvested from the mucosa of the nasal cavity, including the nasoseptal, middle turbinate and inferior turbinate flaps. Regional flaps are harvested from areas adjacent to the nasal cavity and include the extracranial pericranium, temporoparietal fascia and palatal flaps.

Despite the alternatives, the nasoseptal flap remains the first reconstructive choice for most endonasal skull base surgeries. It is a hardy and versatile flap that can reliably reconstruct the majority of ventral cranial base defects [4]. The second choice for most large defects is the extracranial pericranial flap, followed by the extended inferior turbinate flap or temporoparietal fascial flap [5, 6]. The middle turbinate and inferior turbinate flaps are best suited for limited defects of the anterior cranial base and clivus, respectively [7, 8]. There is minimal clinical experience with the palatal flap but it may be suitable for reconstruction of nasopharyngeal and clival defects [9].

Vascular Anatomy

The nasoseptal flap is pedicled upon the posterior septal artery, which is a terminal branch of the sphenopalatine artery [1]. The posterior septal artery corresponds to the posterior branch of the sphenopalatine artery. It ascends after its exit from the sphenopalatine foramen and crosses the anterior wall of the sphenoid sinus between the sphenoid ostium and the arch of the choana to reach the nasal septum. This artery has three branches that contribute to the septal circulation. The superior branch supplies the anterior-superior area of the septum. In this region, the superior branch anatomoses with a branch from the facial artery and with collaterals of the anterior ethmoidal artery. The two inferior branches of the septal artery supply the inferior septal area. The lowest branch enters the incisive canal to reach the hard palate. It anastomoses with an anterior branch of the descending palatine artery. The postero-superior area of the septum is dependent on the posterior ethmoidal artery and its branches [10].

Commonly, the posterior septal artery branches over the anterior wall of the sphenoid sinus before it reaches the nasal septum. In 40% of the cases, there are two branches of the posterior septal artery at the level of the sphenopalatine foramen and in 70% of cases there are two branches at the level of the sphenoid ostium [10, 11]. The mean distance between the ostium and the first branch is 9.3 mm (5–15 mm) [11] (fig. 1a). This anatomical variation explains the potential availability of the septal flap pedicle even in cases of prior sphenoidotomy. A large sphenoidotomy can preclude the use of the nasoseptal flap if a single trunk of the septal artery is present and injured during the procedure. The availability of the vascular pedicle can be assessed with intraoperative Doppler sonography in patients with prior sinus or septal surgery [12].

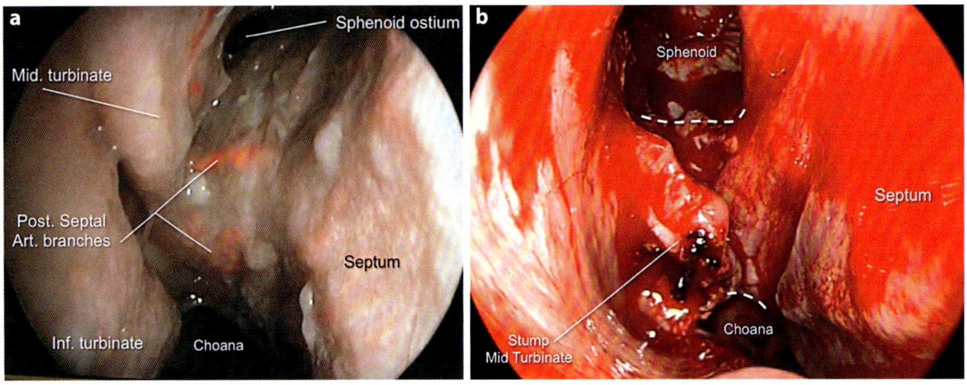

Fig. 1. **a** Endoscopic picture from an anatomical dissection using a 0° endoscope on the right side. The arteries have been injected with red silicone. It is possible to identify two branches of the posterior septal artery crossing the anterior wall of the sphenoid sinus. The pedicle of the nasoseptal flap corresponds to the area between the sphenoid ostium and the arch of the choana. **b** Intraoperative endoscopic picture with a 0° endoscope on the right side. The middle turbinate has been resected without injuring the blood supply to the nasoseptal flap and the stump of the turbinate remains. The sphenoidotomy was completed with preservation of the nasoseptal flap pedicle. The bone behind the pedicle is removed through the contralateral side to avoid injury to the vascular supply to the flap. If needed, the nasoseptal flap can be harvested at the end of the procedure. Mid. turbinate = Middle turbinate; Inf. turbinate = inferior turbinate; Post. septal Art. branches = posterior septal artery branches; Stamp Mid turbinate = stamp of the middle turbinate.

Patient Selection

Most patients are suitable candidates for nasoseptal flap reconstruction. Small defects and low-flow CSF leaks can be repaired effectively using nonvascularized fascial grafts or a free mucosal graft (harvested from a resected middle turbinate or nasal floor). The large size and wide arc of rotation of the septal flap allows the nasoseptal flap to reach defects from the frontal sinus to the lower clivus [13] (fig. 2). The flap can reconstruct suprasellar or infrasellar defects but not both at the same time. If used for a suprasellar defect, it is oriented vertically and can cover a defect that extends from the sella to the frontal sinus and from orbit to orbit. If it is used for an infrasellar defect, it is oriented horizontally and can cover a defect from the sellar floor to the foramen magnum.

The flap can be harvested from either side of the nasal septum, but not both or the septal cartilage is devascularized. The choice of side is dependent on multiple factors: septal deviation, visualization and room for instrumentation, preference of the surgeon, side of the tumor, and side of the dural defect. The flap is usually harvested from the side opposite the tumor extension if preservation of the vascular pedicle would restrict access to the tumor (e.g. sellar tumor with unilateral cavernous sinus extension). Otherwise, the flap is placed on the side of the dural defect to minimize the distance from the pedicle to the defect.

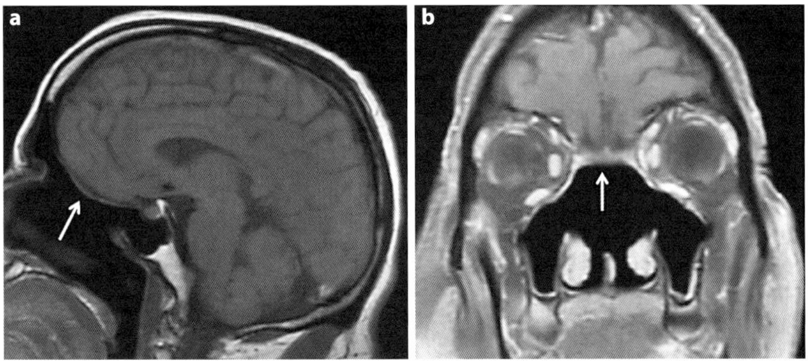

Fig. 2. Postoperative MRI after endoscopic resection of the anterior cranial base for sinonasal malignancy and reconstruction with the nasoseptal flap. **a** Sagittal view shows a well-healed flap (arrow) completely covering the anterior cranial base defect. The frontal sinus and sphenoid sinus remain patent. **b** Coronal view shows the flap (arrow) covering the anterior cranial base defect from orbit to orbit. Note the compete resection of the ethmoidal complex on each side.

A history of prior septal or sinus surgery is not an absolute contraindication to the use of a nasoseptal flap. In patients with a prior septoplasty or trans-sphenoidal approach to the pituitary, the two sides of the septum can be dissected using blunt and sharp dissection without perforation of the flap. If a large sphenoidotomy is present, preservation of a posterior septal artery can be verified using an endoscopic Doppler probe at the beginning of the surgery [12]. The nasoseptal flap can be harvested in patients with limited septal perforations. In such situations, the flap may be designed excluding the perforation from the reconstruction area.

In pediatric patients, differential growth rates of the facial skeleton and cranium limit the use of the nasoseptal flap for reconstruction. Up until the age of 14 years, the septal area is not as large relative to the skull base and it may need to be supplemented with fascial or fat grafts [14]. We have used nasoseptal flaps in patients as young as 4 years of age effectively but the impact of such flaps on subsequent nasal growth is unknown.

Surgical Technique: Harvesting of Flap

Mucosal vasoconstriction is achieved with topical application of oxymetazoline and infiltration of 0.5% xylocaine with 1:200,000 epinephrine around the periphery of the septal flap. Endoscopic visualization with a 0° endoscope is used. Turbinates are lateralized to expose the sphenoethmoidal recess and identify the vascular pedicle that consists of the bridge of mucosa between the sphenoid ostium and the posterior choana. A needle-tip extended monopolar electrocautery device (Bovie Medical Corporation) is used on a low setting (15 W) to make the mucosal incisions. The inferior incision starts at the superior margin of the posterior choana and extends medially along the posterior margin of the vomer toward the nasal floor. The inferior incision continues anteriorly along the junction between the septum and nasal floor to the anterior margin of the septum (junction of septal mucosa and vestibular skin). The superior incision starts at the sphenoid ostium and runs parallel to the skull base approximately 1 cm inferior to the nasal vault (olfactory sulcus). When the incision reaches the level of the anterior attachment of the middle turbinate, it is directed superiorly toward the nasal dorsum to incorporate the most anterior-superior area of the septal mucosa. The incision progresses anteriorly along the nasal dorsum to the anterior edge of the septum. Finally, the superior and inferior incisions are connected with a vertical incision at the mucocutaneous junction. The nasoseptal flap is elevated from the septal cartilage and bone in a subperichondrial and subperiosteal plane to the sphenoid rostrum and anterior face of the sphenoid sinus with preservation of the vascular pedicle [11] (fig. 3, 4) If additional mobility is needed, the flap pedicle can be mobilized to the pterygopalatine space. A middle meatal antrostomy is performed and the sphenopalatine foramen is identified. The overlying bone is carefully removed with a rongeur to expose the contents of the pterygopalatine space. It is important not to injure the sphenopalatine artery. The inferior incision can then be extended from the posterior choana to the lateral nasal wall between the sphenopalatine foramen and the eustachian tube. The flap pedicle is then elevated from the underlying bone of the base of the pterygoids. The limit of lateral elevation is the vidian artery and nerve at the pterygoid canal.

If a wider flap is needed, the mucosa of the floor of the nasal cavity can be incorporated into the nasoseptal flap. In particular, deep transclival defects require a wider flap since the flap is positioned across the defect transversely and may need to extend from the sella to the foramen magnum. In such cases, the inferior incision extends across the nasal floor posteriorly to the inferior meatus. Depending on how much additional mucosa is needed, the mucosa of the inferior meatus can be included. The inferior turbinate should be mobilized medially to provide access to the inferior meatus. The anterior incision extends across the nasal floor anteriorly from the inferior meatus to the anterior septum.

The junction between the nasal floor and the septum is the most difficult area to dissect. The nasopalatine nerve and adherent periosteum along the premaxilla result in tethering of the inferior aspect of the mucosa and must be dissected sharply to avoid tearing the septal flap. It is helpful to dissect the nasal septum and the nasal floor as separate tunnels prior to the dissection along the premaxilla.

In case of sinonasal malignancy, tumor involvement of the nasal septum must be assessed and the area of the nasoseptal flap in proximity to the tumor must be cleared with frozen sections [15]. Incorporation of the mucosa from the septal floor and inferior meatus is usually necessary in such situations to provide adequate width of the flap. If the flap width is still not sufficient or clear oncological margins cannot be confirmed, other reconstructive options should be considered.

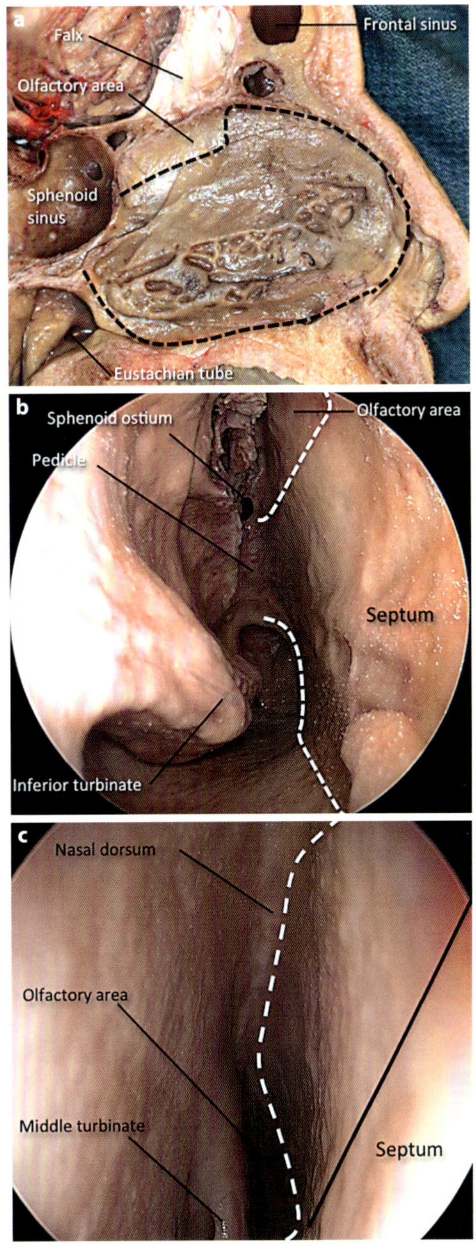

Fig. 3. **a** Anatomical dissection (sagittal view) shows the incisions used to harvest the nasoseptal flap. Note that the superior incision is prolonged superiorly toward the nasal dorsum and the mucosa in the olfactory sulcus is preserved. **b** Cadaveric dissection of a right nasal cavity with a 0° endoscope. Notice the contour along the posterior border of the vomer and the incision is prolonged anteriorly between the septum and the nasal floor. Note the preservation of part of the superior septal mucosa, which contains the olfactory neuroepithelium. **c** Cadaveric dissection with a 0° endoscope looking at nasal dorsum in the right side. Dashed line shows that the superior incision is prolonged toward the nasal dorsum at the level of the middle turbinate attachment. The continuous line demonstrates the cut in the septum if the surgeon does not turn the superior incision toward the nasal dorsum. Note the amount of additional mucosa that can be harvested (area between the dashed line and the continuous line).

After the nasoseptal flap is harvested, the flap can be placed into the nasopharynx for later reconstruction. For sellar and suprasellar approaches, the flap and its pedicle are out of the surgical field. For transclival approaches, the flap is in the way and can be stored in the maxillary sinus after performing a wide middle meatal maxillary antrostomy. A large antrostomy is necessary to prevent ischemic compression of the flap (online suppl. video 1).

OSM

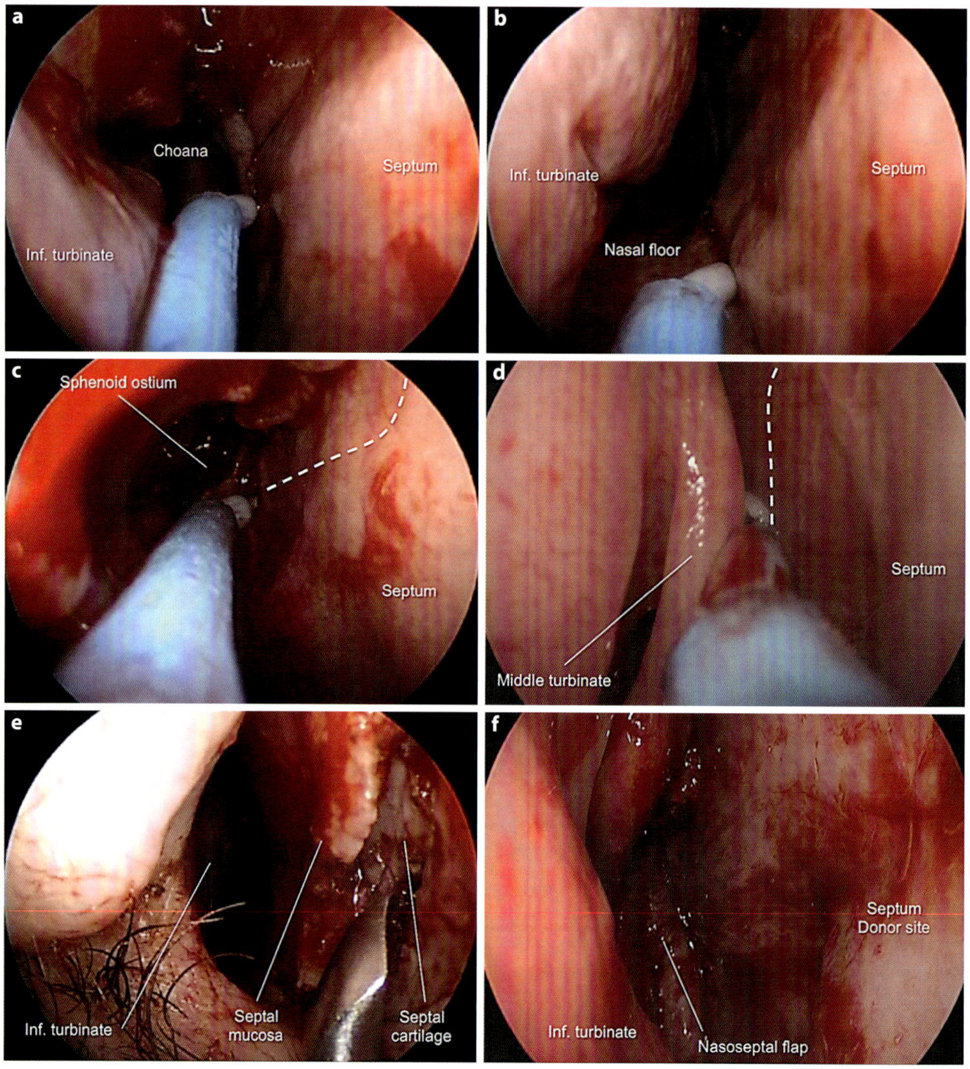

Fig. 4. Intraoperative endoscopic pictures with a 0° endoscope of a right nasoseptal flap being harvested. **a** Posterior incision at the posterior border of the vomer. **b** Inferior incision between the septum and the nasal floor. **c** Superior incision starting from the inferior border of the sphenoid ostium and progressing anteriorly, preserving superiorly the olfactory mucosa. **d** At the level of the attachment of the middle turbinate, the superior incision turns superiorly towards the nasal dorsum. **e** Following an incision at the anterior border of the nasal septum, a Cottle dissector exposes the underlying cartilage. **f** Nasal septum after the flap is harvested in a subperichondrial/subperiosteal plane and transposed into the nasopharynx. Inf. turbinate = Inferior turbinate.

Surgical Technique: Reconstruction of Skull Base Defect

We prefer to perform a two-layer repair of skull base defects using an intradural collagen graft and the extradural nasoseptal flap. A wide variety of materials can be used for the intradural graft: synthetic dural substitute, autologous fascia, cadaveric fascia or subdermal tissue. The inlay graft is interposed between the brain and the dura and extends beyond the margins of the defect [4]. The surrounding dura and bone is cleared of surgical debris and mucosa and the nasoseptal flap is transposed to cover the entire cranial base defect. It is important to mobilize the flap carefully to avoid trauma to the vascular pedicle or tearing of the flap. Torsion or twisting of the vascular pedicle should be minimized and the periosteal surface of the flap should be in contact with the underlying dura and bone. It is imperative to remove all the nasal mucosa in contact with the flap to avoid delayed mucocoeles, especially in the lateral recess of the sphenoid sinus (fig. 5).

The distal portion of the flap corresponding to the anterior-superior nasal septum is the most important part of the flap for reconstruction. This part of the flap is thicker and should be placed at the site of the dural defect. The pedicle of the flap (periosteal surface) should be in contact with tissue to prevent contracture from healing by secondary intention. When the flap pedicle is redundant, it can be folded on itself to provide contact. The medial wall of the orbit, the sphenoid floor and the clival recess are commonly used to support the vascular pedicle for sellar and suprasellar reconstruction. For very anterior defects in the cribriform area, the flap may not reach if it follows the contour of the sphenoid sinus. In such cases, autologous fat grafts can be used to fill the sphenoid sinus and provide a more direct route and extend the reach of the flap.

The surgeon needs to provide meticulous attention to the placement of the flap. This is where most errors occur that result in postoperative CSF leaks. The flap is positioned to cover the entire cranial base defect, taking care to avoid folds along the edges of the flap. If the flap is not large enough to cover the entire defect, it can be supplemented with a turbinate or contralateral septal flap, a nonpermeable fascial graft placed deep to the flap, or autologous fat grafts placed around the periphery of the flap.

The flap needs to be supported externally to counteract intracranial CSF pressures. Small pieces of Surgicel® (Ethicon Inc.) are placed around the periphery of the flap to hold it in position and to prevent tissue glue from undermining the flap. One or two pieces of Gelfoam® (Pfizer) can be place in strategic places over the flap to guarantee the position and contour. Tissue glue is then applied sparingly, using either a synthetic glue (Duraseal™) or fibrin glue (Tisseel®, Baxter). Additional pieces of Gelfoam soaked in thrombin are placed over the surface to provide a cushion between the glue and the packing. For sellar and suprasellar defects, a Foley balloon catheter inflated with saline is preferred since it fits the shape of the cavity and is easy to remove. If the optic nerves/chiasma were exposed during the dissection or are at risk of compression, Merocel® (Medtronic) tampons are used instead. The tampons are also preferred for large defects of the anterior cranial base since they fit the shape of

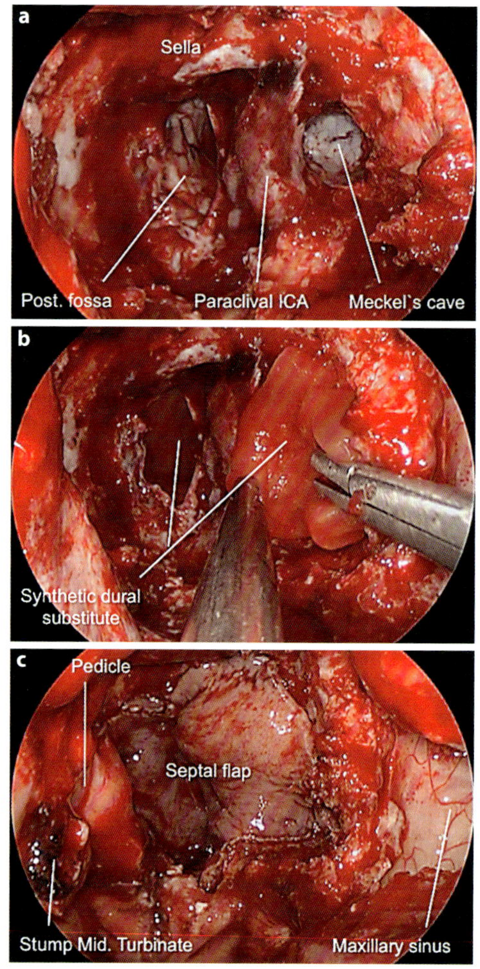

Fig. 5. Endoscopic views during an endoscopic endonasal approach to the middle and posterior cranial fossa for resection of a left trigeminal schwannoma. **a** Endoscopic view with a 0° endoscope showing a defect in the clivus and Meckel's cave (middle fossa). The paraclival carotid artery separates these two areas. **b** Two pieces of synthetic dural substitute are used as inlay grafts. **c** Right nasoseptal flap covering both cranial base defects: clival and Meckel's cave.

the defect better. The packing remains for 5–7 days. A free mucosal graft is harvested from a resected middle turbinate and sutured to the anterior septum to cover the donor site for the flap. This speeds mucosalization of the exposed cartilage and helps minimize long-term postoperative crusting. Silastic septal splints are placed on both sides of the septum to prevent synechiae and promote healing of exposed cartilage, and are removed after 3 weeks.

Modifications of the Nasoseptal Flap: *'Rescue' Nasoseptal Flap*

Harvesting of the nasoseptal flap is associated with increased nasal morbidity. Postoperative crusting, decreased olfaction and intranasal synechiae are more common when the flap is harvested. We would like to avoid these morbidities when a vascularized flap is not necessary for reconstruction. When the anticipated risk of an intraoperative CSF leak is low and the reconstruction can be done without vascularized tissues (e.g. small pituitary adenoma), it is not necessary to elevate a nasoseptal flap at the beginning of the surgery. A free mucosal graft harvested from the middle turbinate is a good reconstructive option in these cases. The need for a flap cannot always be predicted, however, and an unanticipated CSF leak may result. In such cases, the vascular pedicle of the nasoseptal flap can be preserved during the approach and a flap can be harvested at the time of reconstruction. A disadvantage of the 'rescue' flap is the potential trauma to the septal mucosa with passage of instruments during the procedure, especially on the side opposite the endoscope ('blind' side).

An alternative is to elevate a flap in every case but return it to the septum if it is not needed. There may be advantages to performing a nasoseptal flap reconstruction even in the absence of a CSF leak. If there is a large extradural defect with loss of mucosa, a nasoseptal flap provides additional immediate mucosal coverage and promotion of healing with less crusting.

The 'rescue flap' technique consists of a wide sphenoidotomy with preservation of the vascular pedicle of the nasoseptal flap on one side. After the natural ostium of the sphenoid sinus is identified, the ostium is enlarged medially. A short releasing cut (1–2 cm) can be made superiorly from the ostium to the superior septum, corresponding to the superior incision for the standard nasoseptal flap (fig. 1b). The vascular pedicle is then dissected from the anterior face of the sphenoid bone and displaced inferiorly. Completion of the contralateral sphenoidotomy with sacrifice of the contralateral vascular pedicle and communication of both sphenoid sinuses facilitates the removal of bone behind the vascular pedicle [16].

Reuse of the Nasoseptal Flap

One of the advantages of the nasoseptal flap is that it can be reutilized in most cases with benign pathology (recurrent pituitary adenoma, meningioma, craniopharyngioma). The operative notes should be confirmed with the endoscopic examination to identify the side of the pedicle. It is usually quite evident but Doppler sonography can be useful in confirming the side of the vascular pedicle. Nasoseptal flaps should not be reused when there is recurrence of malignant disease or there is obvious invasion by benign tumor.

After confirmation of the side of the pedicle, the margins of the flap are identified and elevated from the bony edges of the defect. A well-defined plane of dissection is usually present between the flap and the underlying pseudodura; it is separated using blunt and sharp dissection. It is important to understand the extent of prior surgery and know what critical structures may have been exposed at the prior surgery in order to avoid injury (fig. 6). The flap is completely mobilized to its vascular pedicle so that it can be displaced from the surgical field.

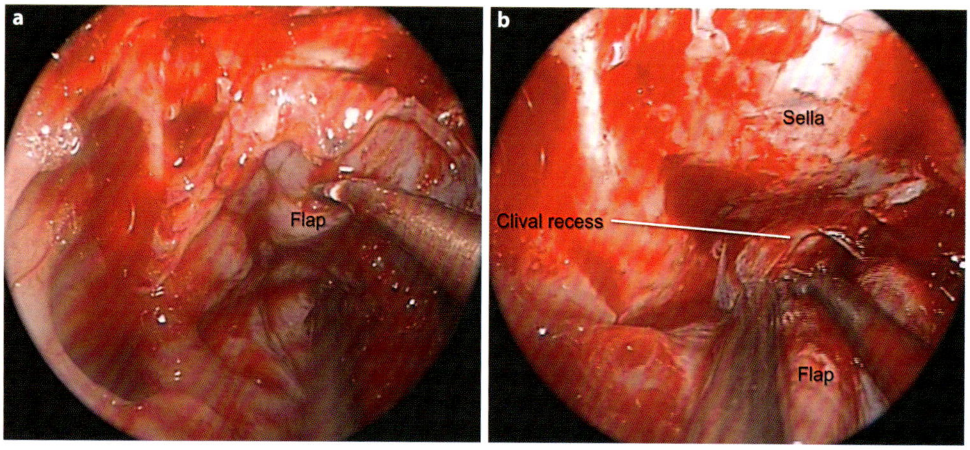

Fig. 6. Intraoperative endoscopic views with a 0° endoscope of a reoperation of a patient with a recurrent pituitary tumor. The nasoseptal flap was used for reconstruction during the first surgery and the flap is harvested from the cranial base defect. **a** The edges of the flap are identified and careful dissection of the flap from underlying bone and dura is performed to avoid injury to the flap or to the underlying neurovascular structures. **b** Dissection of the flap from the neodura in the clival region.

Following the resection, the flap retains the contour of the original defect. This facilitates the positioning of the flap during the reconstruction but can be problematic if the shape or size of the defect is altered. In such cases, the reconstruction may need to be supplemented with fascial or fat grafts.

Avoiding and Managing Intraoperative Complications

Complications associated with nasoseptal flap reconstruction can be divided into those associated with harvesting of the flap and those associated with reconstruction of the defect. Patients with history of prior septoplasty or prior transseptal transsphenoidal surgery are at greater risk of laceration or tearing of the flap during dissection. A severely deviated septum or sharp septal spur also makes dissection difficult. If a laceration of the flap happens during harvesting, the surgeon can try to orient the flap so that the hole does not overlay the cranial base defect. If the laceration coincides with the cranial base defect, a layer of collagen matrix can be placed just underneath the flap. On the mucosal side, the laceration can be sutured or a small plug of fat graft can be placed through the hole.

Intraoperative loss of the blood supply to the flap is extremely rare. Fortunately, the vascular anatomy of the pedicle of the flap is very predictable and tolerates manipulation. Nevertheless, the pedicle should be protected when drilling or using a microdebrider. Excessive electrocautery in proximity to the pedicle should be avoided. If the flap is stored in the maxillary sinus during surgery, the antrostomy should be large enough to avoid vascular compression. Finally, excessive torsion of the pedicle should be avoided. If there is concern about injury to the pedicle, the vascular supply can be checked with Doppler sonography. Even if the vascularity of the flap is compromised, the flap can be used for reconstruction as a nonvascularized mucosal graft.

The most common reason for failure of a reconstruction with resultant CSF leak is a flap that is too small for the defect. Techniques to maximize the size of the flap include incorporation of the anterior-superior septal mucosa subjacent to the nasal bones and the mucosa of the nasal floor. In order to avoid tension on the closure and extend the reach of the flap, the pedicle should be fully mobilized. Removal of the sphenoid floor and filling the sphenoid sinus with fat grafts will further extend the reach of the flap.

Uncommon errors for failure include placement of the flap with the mucosal surface in contact with the dura and bone instead of the periosteal surface and delayed necrosis. If the flap pedicle is not in contact with bone, contraction of the flap may occur with healing, resulting in retraction of the flap and disruption of the repair.

Septum Donor Site

The exposed septal cartilage is great source of crusting during the postoperative period. It takes weeks for mucosalization of the exposed cartilage. The healing of the mucosa over the septum is accelerated with placement of a free mucosal graft on the exposed cartilage. Usually the mucosal graft is harvested from a resected middle turbinate and sutured to the anterior septum. Silastic splints are sutured to the septum for 3 weeks to promote healing of the graft and mucosalization of exposed areas of cartilage (fig. 7). Another option for coverage of the exposed cartilage is to reverse the contralateral posterior septal mucosa and suture it to the cartilage on the side of the flap [17]. This decreases the area of exposed cartilage for mucosalization but requires additional resection of posterior septum for the mucosal flap to cover the anterior septum.

Conclusion

The nasoseptal flap satisfies most reconstructive needs of endoscopic endonasal surgery of the skull base. The reliable vascular supply promotes rapid healing and is an effective barrier for the prevention of CSF leaks (fig. 8). The nasoseptal flap has a good arc of rotation and can reach defects of the ventral skull base from the frontal sinuses to the foramen magnum. It remains our first choice for large dural defects and has dramatically decreased the risk of a postoperative CSF leak to less than 5%.

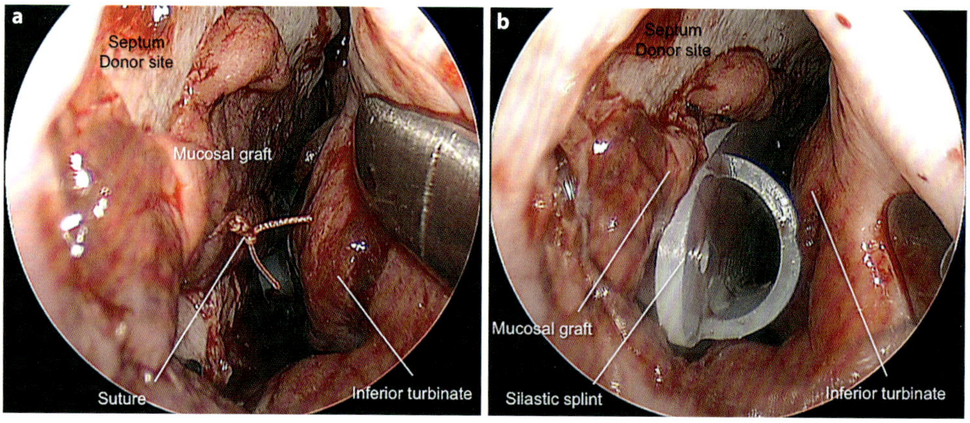

Fig. 7. Intraoperative endoscopic views with a 0° endoscope demonstrates the free mucosal graft used to accelerate the healing at the septum donor site. **a** The middle turbinate mucosal graft is placed to cover the cartilage exposed in the septum donor site anteriorly. The graft is sutured to the septal cartilage with absorbable suture. **b** Silastic splints with airway tunnels provide compression of the graft against the cartilage.

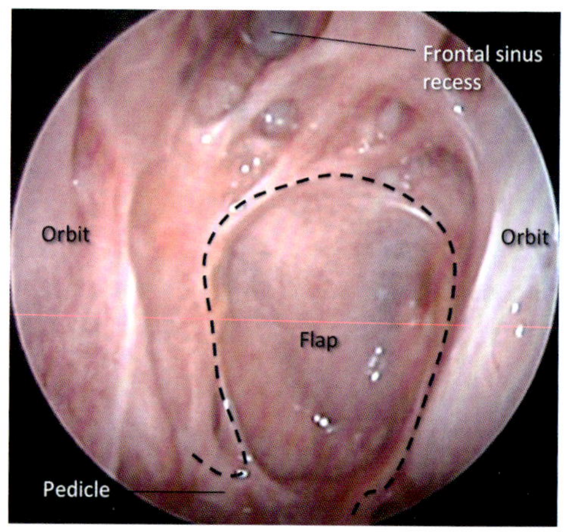

Fig. 8. Endoscopic view with a 0° endoscope of a patient who had endoscopic resection of the anterior cranial base and reconstruction with a nasoseptal flap. The flap is healed and completely covers the anterior cranial base.

References

1 Hadad G, Bassagasteguy L, Carrau RL, Mataza JC, Kassam A, Snyderman CH, Mintz A: A novel reconstructive technique after endoscopic expanded endonasal approaches: vascular pedicle nasoseptal flap. Laryngoscope 2006;116:1882–1886.

2 Zanation AM, Carrau RL, Snyderman CH, Germanwala AV, Gardner PA, Prevedello DM, Kassam AB: Nasoseptal flap reconstruction of high flow intraoperative cerebral spinal fluid leaks during endoscopic skull base surgery. Am J Rhinol Allergy 2009;23:518–521.

3 Harvey RJ, Parmar P, Sacks R, Zanation AM: Endoscopic skull base reconstruction of large dural defects: a systematic review of published evidence. Laryngoscope 2012;122:452–459.

4 Snyderman CH, Kassam AB, Carrau R, Mintz A: Endoscopic reconstruction of cranial base defects following endonasal skull base surgery. Skull Base 2007;17:73–78.
5 Fortes FS, Carrau RL, Snyderman CH, Kassam A, Prevedello D, Vescan A, Mintz A, Gardner P: Transpterygoid transposition of a temporoparietal fascia flap: a new method for skull base reconstruction after endoscopic expanded endonasal approaches. Laryngoscope 2007;117:970–976.
6 Zanation AM, Snyderman CH, Carrau RL, Kassam AB, Gardner PA, Prevedello DM: Minimally invasive endoscopic pericranial flap: a new method for endonasal skull base reconstruction. Laryngoscope 2009;119:13–18.
7 Prevedello DM, Barges-Coll J, Fernandez-Miranda JC, Morera V, Jacobson D, Madhok R, dos Santos MC, Zanation A, Snyderman CH, Gardner P, et al: Middle turbinate flap for skull base reconstruction: cadaveric feasibility study. Laryngoscope 2009;119: 2094–2098.
8 Fortes FS, Carrau RL, Snyderman CH, Prevedello D, Vescan A, Mintz A, Gardner P, Kassam AB: The posterior pedicle inferior turbinate flap: a new vascularized flap for skull base reconstruction. Laryngoscope 2007;117:1329–1332.
9 Oliver CL, Hackman TG, Carrau RL, Snyderman CH, Kassam AB, Prevedello DM, Gardner P: Palatal flap modifications allow pedicled reconstruction of the skull base. Laryngoscope 2008;118:2102–2106.
10 Babin E, Moreau S, de Rugy MG, Delmas P, Valdazo A, Bequignon A: Anatomic variations of the arteries of the nasal fossa. Otolaryngol Head Neck Surg 2003;128:236–239.
11 Pinheiro-Neto CD, Ramos HF, Peris-Celda M, Fernandez-Miranda JC, Gardner PA, Snyderman CH, Sennes LU: Study of the nasoseptal flap for endoscopic anterior cranial base reconstruction. Laryngoscope 2011;121:2514–2520.
12 Pinheiro-Neto CD, Carrau RL, Prevedello DM, Fernandez-Miranda JC, Snyderman CS, Gardner PA, Kassam AB: Use of acoustic Doppler sonography to ascertain the feasibility of the pedicled nasoseptal flap after prior bilateral sphenoidotomy. Laryngoscope 2010;120:1798–1801.
13 Pinheiro-Neto CD, Prevedello DM, Carrau RL, Snyderman CH, Mintz A, Gardner P, Kassam A: Improving the design of the pedicled nasoseptal flap for skull base reconstruction: a radioanatomic study. Laryngoscope 2007;117:1560–1569.
14 Shah RN, Surowitz JB, Patel MR, Huang BY, Snyderman CH, Carrau RL, Kassam AB, Germanwala AV, Zanation AM: Endoscopic pedicled nasoseptal flap reconstruction for pediatric skull base defects. Laryngoscope 2009;119:1067–1075.
15 Snyderman CH, Carrau RL, Kassam AB, Zanation A, Prevedello D, Gardner P, Mintz A: Endoscopic skull base surgery: principles of endonasal oncological surgery. J Surg Oncol 2008;97:658–664.
16 Rivera-Serrano CM, Snyderman CH, Gardner P, Prevedello D, Wheless S, Kassam AB, Carrau RL, Germanwala A, Zanation A: Nasoseptal 'rescue' flap: a novel modification of the nasoseptal flap technique for pituitary surgery. Laryngoscope 2011; 121:990–993.
17 Caicedo-Granados E, Carrau R, Snyderman CH, Prevedello D, Fernandez-Miranda J, Gardner P, Kassam A: Reverse rotation flap for reconstruction of donor site after vascular pedicled nasoseptal flap in skull base surgery. Laryngoscope 2010;120:1550–1552.

Carl H. Snyderman, MD, MBA
Departments of Otolaryngology and Neurological Surgery,
University of Pittsburgh School of Medicine
200 Lothrop Street, EEI 500, Pittsburgh, PA 15213 (USA)
E-Mail snydermanch@upmc.edu

OSM

Bleier BS (ed): Comprehensive Techniques in CSF Leak Repair and Skull Base Reconstruction.
Adv Otorhinolaryngol. Basel, Karger, 2013, vol 74, pp 56–63 (DOI: 10.1159/000342275)

Sphenopalatine-Derived Pedicled Flaps

Jeffrey D. Suh[a] · Alexander G. Chiu[b]

[a]Department of Head and Neck Surgery, UCLA School of Medicine, Los Angeles, Calif., and
[b]Division of Otolaryngology, Head and Neck Surgery, University of Arizona, Tucson, Ariz., USA

Abstract

A variety of local flaps based on the sphenopalatine artery are available for endoscopic skull base reconstruction including the nasoseptal flap, middle turbinate flap, and posteriorly based inferior turbinate flap. This section will describe the alternatives to the nasoseptal flap, emphasizing the relevant anatomy, indications and technical aspects involved with their use.

The success of endoscopic skull base surgery can be largely attributed to advances in skull base reconstruction. The Hadad-Bassagasteguy flap (HBF), as described by the group at the University of Pittsburgh in 2006, has allowed surgeons to address almost any size defect of the ventral skull base, resulting in a sharp decrease in postoperative CSF leaks [1]. It is this ability to reliably separate the cranial cavity from the sinonasal tract that has allowed the field of endoscopic skull base surgery to flourish. Since then, others have described alternatives to the nasoseptal flap. This chapter will focus on the sphenopalatine artery-derived flaps, specifically from the middle and inferior turbinates, emphasizing technique, indications, limitations and results.

Vascularized versus Nonvascularized Grafts

Free tissue grafts have a long history in head and neck reconstruction. For example, buccal mucosal grafts are used successfully in tracheal and laryngeal reconstruction and after conjunctivodacryocystorhinostomy [2]. Skin grafts are successfully used to line orbital exenteration defects. In 1981, Wigand [3] introduced the concept of single-layer closure of small-to-medium-sized CSF leaks using free mucosal grafts. Currently, a variety of regional donor sites are utilized in endonasal endoscopic repair, including mucosa from the nasal septum, nasal floor and turbinates [4–6]. Distant donor sites include the hard palate mucosa, fascia lata and temporalis fascia. Also, a variety of allografts can be used with success [6, 7].

The pathophysiology of wound healing of free grafts was not well understood until a series of elegant experiments by Hoseman et al. [8] in 1999. Using a rabbit model, they created 2-mm skull base defects and CSF leaks which were then repaired with free mucosal grafts from the nasal septum. The rabbits were then sacrificed at different postoperative periods, ranging from 2 days to 12 weeks, and the defects analyzed by light microscopy. After 48 h, the respiratory epithelium of the mucosal graft disappeared. Blood vessels in the graft were congested, and the lamina propria was infiltrated by inflammatory cells. After 4 days, the cellular infiltration continued to increase. After 1 week, the grafts increased in thickness. After 2 weeks, the original lamina propria changed and was replaced largely by immature fibrous tissue. Capillaries from the skull base bone were seen invading the graft, and the bone showed signs of osteogenesis. After 3 weeks, capillaries were found in the graft. Osteogenesis peaked at this time, closing the skull base defect completely with immature bone. After 8 weeks, the outline of the graft could no longer be identified by microscopy. The grafts were transformed into scar tissue with intercalated bony trabeculae. The authors reported a 20% reduction in the size of the grafts, which started to occur in the first days after surgery. For this reason, they recommended that harvested grafts be 25% larger than the defect.

In general, the advantages of free tissue grafts for endoscopic skull base reconstruction include relatively easy harvest and insetting with minimal donor site morbidity. Disadvantages of free grafts include the risk of partial or complete tissue necrosis causing delayed CSF leaks. Risk factors for graft failure include: a history of prior surgery, radiation, large dural defects, benign intracranial hypertension, high-flow CSF leaks, and noncompliance with postoperative instructions [6, 9, 10]. However, as illustrated in the chapter by Metson [this vol.], free grafts have been used with a high rate of success for a majority of small-to-medium-sized defects.

In contrast, large surgical defects of the skull base after craniofacial resection are often repaired using regional vascularized flaps such as the pericranial and temporoparietal flaps. These flaps are quite successful in repairing defects of the anterior skull base, with a 95% overall success rate [11]. Compared to free tissue grafts, vascularized flaps are thought to be more reliable and heal faster, especially in the setting of prior radiation or surgery. The blood supply to local flaps is via the preservation of a vascular pedicle at the base of the flap, which maintains continuity with the donor site following transfer. The most common pedicled flap used in endoscopic skull base surgery is the nasoseptal flap, which is covered in detail in the chapter by Snyderman and Pinheiro-Neto [this vol.]. This chapter will focus on two other pedicled flaps based on the sphenopalatine artery, the inferior and middle turbinate flaps.

Technical Pearls

Our preoperative evaluation consists of a thorough history and physical examination, nasal endoscopic examination, and radiographic imaging. During surgery, leak site identification, site preparation, accurate graft placement and postoperative management are critical to success.

Intrathecal fluorescein has been advocated to assist in the localization of skull base defects. Multiple large series [12–15] demonstrate an acceptable safety profile at low doses (<50 mg). We have modified our practice over the years, and use lumbar drains now only infrequently in our skull base surgeries. Only rarely are the lumbar drains left in postoperatively except in cases of benign intracranial hypertension. When used, we follow a protocol previously described by Schlosser et al. [16] at the University of Pennsylvania. Lumbar drains are opened at the time of graft placement, and the height of the collection chamber is adjusted to maintain drainage at 5–10 ml/h. We typically remove the drain on the second postoperative day. Before removing the drain, it is our practice to measure ICP. To do this, on the morning of postoperative day 2 we turn the 3-way stopcock attached to the lumbar drain system to the off position to clamp the drain. 3–4 h are allowed to pass to replenish the patient's CSF volume; then a standardized pressure transducer is connected to one of the ports. The transducer is zeroed with the patient lying perfectly flat and at rest. The pressure monitor displays CSF pressure measurements and dynamic waveforms. A Queckenstedt test is performed to confirm that the pressure measured through the lumbar drain catheter in the lumbar cistern is actually recording the intracranial CSF pressure. This test is performed by compressing both internal jugular veins to increase ICP. It is done while observing the waveform and pressure measurements and typically results in an increase in ICP of 15–20 cm H_2O within 5–10 s and an increased amplitude of the wave form. CSF pressure is then recorded. If the CSF pressure is significantly elevated, consideration is made for placing the patient on acetazolamide postoperatively.

Inferior Turbinate Flap

This flap was first described in 2007 by Fortes et al. [17]. The pedicle of this flap is based on the inferior turbinate artery, a terminal branch of the PLNA, which is a branch of the sphenoplatine artery (SPA). Indications for this flap include those patients in which the HBF is not available, such as when there is a septectomy defect, tumor involvement of the septum, or when the blood supply has been interrupted such as after a wide sphenoidotomy. The most common use for this flap is for clival and sellar reconstruction (table 1). While not based on the SPA, another inferior turbinate flap that has been recently described for endoscopic skull base reconstruction

Table 1. Sphenopalatine-based flaps for skull base reconstruction

Defect	Nasoseptal	Inferior turbinate	Middle turbinate
Sella	yes	yes	yes
Tuberculum/planum	yes	no	yes
Fovea ethmoidalis	yes	no	yes
Cribriform plate	yes	no	no
Clivus	yes	yes	no

Adapted from Prevedello et al. [19].

is the anteriorly based inferior turbinate flap [18]. This flap is based off of the anterior lateral branches of the anterior ethmoid artery, and the lateral nasal artery, which is a branch of the facial artery. The authors report use of this flap in 7 patients in patients where the nasoseptal flap was unavailable or in conjunction with it. In their series, the anteriorly based inferior turbinate flap was able to cover defects in the posterior table to the frontal sinus as well as the cribriform plate. Follow-up in this study demonstrated a 100% flap survival.

Anatomy

This posteriorly based inferior turbinate flap is based on the SPA, one of the terminal branches of the internal maxillary artery. The SPA enters the nasal cavity via the sphenopalatine foramen. It supplies most of the lateral nasal wall, including the inferior and middle turbinates from the posterior lateral nasal artery, also known as the nasolateral artery. A descending arterial trunk enters the inferior turbinate on the superior aspect of its lateral attachment, 1.0–1.5 cm from its posterior tip. The vasculature of the artery within the turbinate can be quite variable [17]. Most commonly, the artery runs from posterior to anterior (mean 1.2 cm) before piercing the bone and soft tissue and splitting off into two to six branches.

Surgical Technique (online suppl. video 1)

Our technique is similar to that described by Fortes et al. [17] (fig. 1). The flap is harvested prior to reconstruction with the skull base defect site already prepared. We usually employ a multilayer closure with the flap representing the overlay graft. It is best to identify the SPA as it exits from the sphenopalatine foramen (as in an endoscopic sphenopalatine artery ligation) and then follow the PLN artery inferiorly into the inferior turbinate to avoid injury to the pedicle. Two parallel incisions are then made. The superior incision is just above the inferior turbinate (over the maxillary sinus fontanelle) and the inferior one at the inferior margin of the turbinate. A vertical incision is then made at the head of the inferior turbinate over the lateral nasal

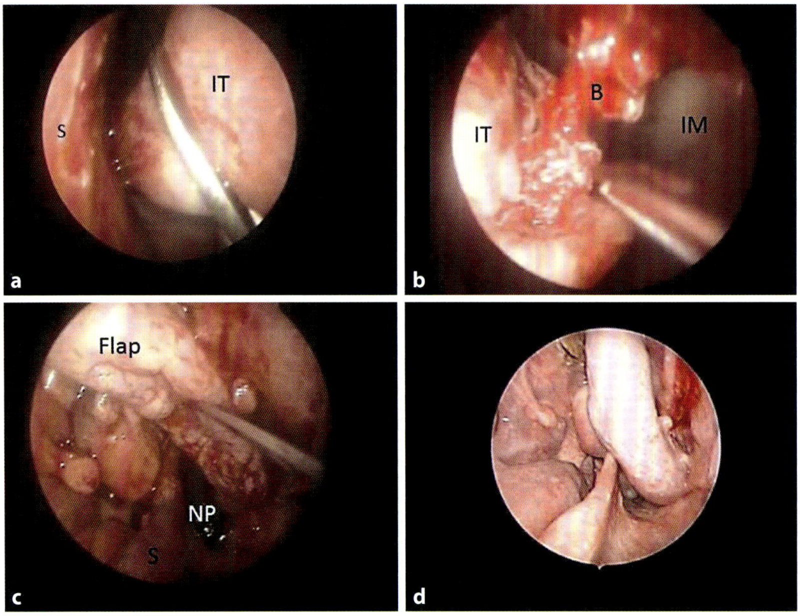

Fig. 1. Inferior turbinate flap. **a** Mucosal incision made on left inferior turbinate. **b** Dissection of mucosa off of turbinate bone. **c** Intraoperative placement of the flap. **d** Inferior turbinate flap at 3-months after surgery. S = Septum; IT = inferior turbinate; B = turbinate bone; IM = inferior meatus; NP = nasopharynx; Flap = elevated inferior turbinate flap.

wall. Incisions can be made with a combination of electrocautery, turbinate scissors, or otologic straight and curved blades depending on preference. A freer or suction elevator is used to lift the mucoperiosteum off of the turbinate bone from anterior to posterior. If more width is required, the flap can be extended further in the inferior meatus, or higher on the lateral nasal wall. Care must be made posteriorly to not injure the vascular pedicle. Once harvested, the flap is gently unrolled and mobilized to cover the skull base defect.

Outcomes

Complications are unusual with the IT flap harvest or inset. Crusting over the lateral nasal wall and frequent debridement until re-mucosalization is complete, usually in 3–4 weeks. The posteriorly based IT flap provides a reliable vascular pedicle with a good arc of rotation best for repairing defects of the clivus and sella. In our experience, the flap cannot reach the anterior skull base. Bilateral flaps can be harvested for additional width to cover larger defects. Relative contraindications to this flap include cases with a history of prior turbinectomy, large defects, or skull base defects located anteriorly due to inadequate pedicle length. Chen et al. [19] reported that use of the IT flap after endoscopic nasopharyngectomy resulted in severe xerosis in some of his patients as well as a hypernasal voice.

Middle Turbinate Flap

The middle turbinate has been evaluated as another pedicled flap for skull base repair [20, 21]. Harvesting this flap requires precise endoscopic dissection to be successful. In a recent series, this flap was used successfully for a variety of sellar defects with both low-flow and high-flow CSF leaks [21]. This flap can be considered either as an alternative to the nasoseptal flap when it is not available or as a first choice for select skull base defects.

Anatomy
The middle turbinate is associated with many functions of the nasal cavity, including olfaction, humidification, lubrication of the upper airways, regulation of airflow and temperature, and filtration [22]. The middle turbinate is also an important landmark of the nasal cavity during endoscopic sinus surgery. Cadaveric studies demonstrate that the middle turbinate is 18.5 ± 4.1 mm long, and 10.9 ± 2.4 mm high [23]. The middle turbinate bone attaches to two areas: the superior attachment to the lateral aspect of the cribriform plate, and the lateral attachment, the basal lamella, to the thin lamina papyracea of the lateral ethmoidal wall. Posteriorly, the basal lamella curves superiorly and becomes oriented in the coronal plane, behind the ethmoidal bulla, thereby separating the anterior and posterior ethmoidal air cells. Branches of the SPA enter its posterior aspect.

Surgical Technique (online suppl. video 2)
This technique has been described well by both Prevedello et al. [20] and Simal et al. [21]. After infiltration, the anterior middle turbinate is incised vertically (fig. 2). The mucosa is elevated off the turbinate bone posteriorly and inferiorly to the posterior pedicle. In our experience it is important that the turbinate not be destabilized prior to harvest, because the mucosa is often very adherent to the bone and can require some pressure to elevate. It is critical that the superior cut (fig. 2c) remain below the level of the cribriform plate to avoid a CSF leak. Then the bone is removed piecemeal. A horizontal cut is made at the axilla, superiorly, using sharp instruments. Finally, the flap is mobilized to cover the defect and secured in place using packing. In one study, the mean time required to prepare the middle turbinate flap was 27 min [21].

Outcomes
The middle turbinate flap is best used for small-to-medium-sized defects of the sella, planum or cribriform plate (table 1). In a cadaveric study, the mean length of the flap was 4 cm, with an average width of 2.8 cm [20]. The flap was considered to be large enough in 10/12 sides to cover a sellar defect. The authors concluded that the flap would be more suitable for covering defects of the planum and fovea ethmoidalis. Chen et al. [19] used the MT flap in 18 patients with recurrent nasopharyngeal

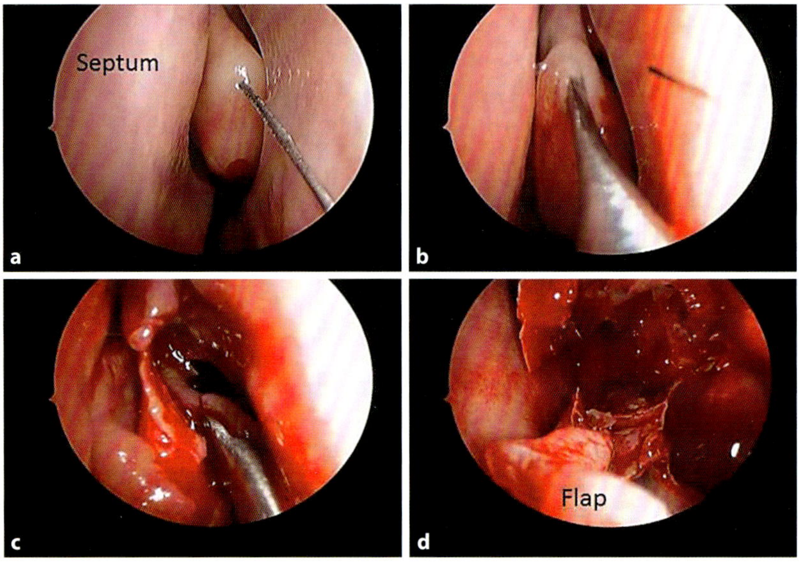

Fig. 2. Middle turbinate flap. **a** Left middle turbinate is infiltrated. **b** A sickle knife is used to make a vertical incision in the turbinate. **c** The mucosa is elevated off of the turbinate bone. **d** The flap is freed completely and ready for placement.

carcinoma after endoscopic nasopharyngectomy. They found that the pedicled flaps reduced crusting and odor by covering the exposed, radiated bone, better than free mucosal grafts or skin grafts.

Postoperative Management

We limit patients to light activity without heavy lifting or straining for 4–6 weeks postoperatively after skull base reconstruction. All patients are placed on stool softeners after surgery. Patients with elevated intracranial pressure are occasionally placed on acetazolamide to lower intracranial pressure or, in some cases, require ventriculoperitoneal shunting. Antistaphylococcal antibiotics are prescribed for 2 weeks after surgery. Regular debridements are performed to remove crusts and clots until the sinonasal mucosa normalizes.

Conclusion

The inferior and middle turbinate flaps represent alternatives to the nasoseptal flap for skull base reconstruction. While both flaps can be technically challenging to harvest, they have been used successfully to repair a variety of defects with little donor site morbidity.

References

1 Hadad G, Bassagasteguy L, Carrau RL, et al: A novel reconstructive technique after endoscopic expanded endonasal approaches: vascular pedicle nasoseptal flap. Laryngoscope 2006;116:1882–1886.

2 Leone CR Jr: Conjunctivodacryocystorhinostomy with buccal mucosal graft. Arch Ophthalmol 1995; 113:113–115.

3 Wigand ME: Transnasal ethmoidectomy under endoscopical control. Rhinology 1981;19:7–15.

4 Lindstom DR, Toohill RJ, Loehrl TA, Smith TL: Management of cerebrospinal fluid rhinorrhea: the Medical College of Wisconsin Experience. Laryngoscope 2004;114:969–974.

5 Suh JD, Ramakrishnan VR, DeConde AS: Nasal floor free mucosal graft for skull base reconstruction and cerebrospinal fluid leak repair. Ann Otol Rhinol Laryngol 2012;121:91–95.

6 Snyderman CH, Kassam AB, Carrau R, et al: Endoscopic reconstruction of cranial base defects following endonasal skull base surgery. Skull Base 2007;17:73–78.

7 Germani RM, Vivero R, Herzallah IR, et al: Endoscopic reconstruction of large anterior skull base defects using acellular dermal allograft. Am J Rhinol 2007;21:615–618.

8 Hosemann W, Goede U, Sauer M: Wound healing of mucosal autografts for frontal cerebrospinal fluid leaks – clinical and experimental investigations. Rhinology 1999;3:108–112.

9 Hegazy HM, Carrau RL, Snyderman CH, et al: Transnasal endoscopic repair of cerebrospinal fluid rhinorrhea: a meta-analysis. Laryngoscope 2000; 110:1166–1172.

10 Woodworth BA, Prince A, Chiu AG, et al: Spontaneous CSF leaks: a paradigm for definitive repair and management of intracranial hypertension. Otolaryngol Head Neck Surg 2008;138:715–720.

11 Price JC, Loury M, Carson B, et al: The pericranial flap for reconstruction of anterior skull base defects. Laryngoscope 1988;98:1159–1164.

12 Felisati G, Bianchi A, Lozza P, et al: Italian multicentre study on intrathecal fluorescein for craniosinusal fistulae. Acta Otorhinolaryngol Ital 2008; 28:159–163.

13 Keerl R, Weber RK, Draf W, et al: Use of sodium fluorescein solution for detection of cerebrospinal fluid fistulas: an analysis of 420 administrations and reported complications in Europe and the United States. Laryngoscope 2004;114:266–272.

14 Placantonakis DG, Tabaee A, Anand VK, et al: Safety of low-dose intrathecal fluorescein in endoscopic cranial base surgery. Neurosurgery 2007;61: 161–165.

15 Seth R, Rajasekaran K, Benninger MS, et al: The utility of intrathecal fluorescein in cerebrospinal fluid leak repair. Otolaryngol Head Neck Surg 2010; 143:626–632.

16 Schlosser RJ, Wilensky EM, Grady MS, et al: Cerebrospinal fluid pressure monitoring after repair of cerebrospinal fluid leaks. Otolaryngol Head Neck Surg 2004;130:443–448.

17 Fortes FS, Carrau RL, Snyderman CH, et al: The posterior pedicle inferior turbinate flap: a new vascularized flap for skull base reconstruction. Laryngoscope 2007;117:1329–1332.

18 Gil Z, Margalit N: Anteriorly based inferior turbinate flap for endoscopic skull base reconstruction. Otolaryngol Head Neck Surg 2012;146:842–847.

19 Chen MY, Hua YJ, Wan XB, et al: A posteriorly pedicled middle turbinate mucoperiosteal flap resurfacing nasopharynx after endoscopic nasopharyngectomy for recurrent nasopharyngeal carcinoma. Otolaryngol Head Neck Surg 2012;146: 409–411.

20 Prevedello DM, Barges-Coll J, Fernandez-Miranda JC, et al: Middle turbinate flap for skull base reconstruction: cadaveric feasibility study. Laryngoscope 2009;119:2094–2098.

21 Simal Julián JA, Miranda Lloret P, Cárdenas Ruiz-Valdepeñas E, et al: Middle turbinate vascularized flap for skull base reconstruction after an expanded endonasal approach. Acta Neurochir (Wien) 2011; 153:1827–1832.

22 Swanson PB, Lanza DC, Vining EM, Kennedy DW: The effect of middle turbinate resection upon the frontal sinus. Am J Rhinol 1995;9:191–195.

23 Lee HY, Kim C-H, Kim JY, Kim JK, Song MH, Yang HJ, Kim K-S, Chung IH, Lee J-G, Yoon J-H: Surgical anatomy of the middle turbinate. Clin Anat 2006;19: 493–496.

Jeffrey D. Suh, MD
Department of Head and Neck Surgery, UCLA School of Medicine
200 UCLA Medical Plaza Suite 550
Los Angeles, CA 90095 (USA)
E-Mail jeffsuh@mednet.ucla.edu

Bleier BS (ed): Comprehensive Techniques in CSF Leak Repair and Skull Base Reconstruction.
Adv Otorhinolaryngol. Basel, Karger, 2013, vol 74, pp 64–70 (DOI: 10.1159/000342281)

Anteriorly Based Pedicled Flaps for Skull Base Reconstruction

Joshua C. Meier · Benjamin S. Bleier

Department of Otology and Laryngology, Massachusetts Eye and Ear Infirmary, Harvard Medical School, Boston, Mass., USA

Abstract

The expansion of endoscopic skull base surgery has resulted in the creation of large defects that must be repaired to ensure separation of the cranial vault from the nasal cavity. The workhorse of anterior skull base reconstruction remains the nasoseptal or Hadad-Bassagasteguy flap. Despite its success, the nasoseptal flap is limited in its ability to reach extremely anterior defects including those involving the frontal break, posterior frontal table, and anterior cribiform plate. Alternative approaches utilizing anteriorly pedicled flaps have been described which exploit the vascular supply to the anterior septum and lateral nasal wall. The diminutive nature of the anterior septal blood supply has led to the elaboration of a bipedicled anterior septal flap which is capable of reliable reconstruction of both the frontal beak as well as posterior frontal table defects. Similarly anteriorly based inferior turbinate flaps pedicled largely on anterior ethmoid arborizations have been used successfully to reconstruct anterior defects. In light of the limitations of the nasoseptal flap, the addition of anteriorly pedicled flaps to the clinical armamentarium offers the opportunity to provide vascularized mucosal coverage of virtually any region of the skull base which can be reached endoscopically.

Recent advances in skull base surgical techniques and instrumentation have resulted in an expansion in the size and location of lesions that are amenable to endoscopic resection. As a result new techniques have been developed to repair the resultant skull base defects. While small defects may be successfully managed by a variety of methods including the use of free grafts, the large defects associated with endoscopic extended endonasal approaches require more robust reconstruction. Techniques previously described included Alloderm, avascular mucosal free grafts, and randomly based flaps however post-operative CSF leak rates were unacceptably high. A major advance in recent years has been the introduction of vascularized septal and lateral wall flaps. Outcomes studies using the nasoseptal or Hadad-Bassagasteguy flap (HBF) demonstrated superior postoperative CSF leak rates relative to prior free graft techniques [1]. Patel et al. [2] in 2010 described the North Carolina and Pittsburgh experiences with use of vascularized flaps for reconstruction of skull base defects. Their use of the nasoseptal flap resulted in a postoperative leak rate of 4% in a prospective series of 150 patients. Based on this success, further descriptions of pedicled inferior and middle turbinate flaps were subsequently developed adding additional options to the armamentarium for skull base repair [3, 4].

Limitations of Posteriorly Pedicled Flaps

The majority of recently described pedicled flaps are derived from the sphenopalatine arterial (SPA) arborizations due to its large-bore and predictable terminal branches. While these flaps may subtend a large mucosal surface area, they are necessarily limited in their reach due to the posterior location of the sphenopalatine artery. The posterior tethering by the sphenopalatine foramen coupled with the inherent limits of flap length result in difficulty with providing adequate coverage for extremely anterior sites such as anterior cribiform and posterior frontal table defects. While some groups have reported the successful use of septal flaps for limited frontal sinus defects these repairs may be tenuous and postoperative flap retraction can result in failure at the repair site. Virgin et al. [5] in 2012 reported the University of Alabama experience using a pedicled nasoseptal flap to repair frontal sinus defects. They retrospectively reviewed a series of patients who underwent nasoseptal flap repair for frontal sinus defects. They deemed 14 of 15 patients as successful repairs, however, 1 patient developed pneumocephalus requiring a craniotomy for definitive repair. Of note, in this study the ability to reach lateral frontal sinus defects required draping of the flap over the lamina to achieve adequate reach.

While reproducible coverage of large posterior frontal defects remains a challenge, there is additional evidence that vascularized mucosal coverage of the denuded frontal beak and anterior frontal table following endoscopic modified Lothrop procedures (EMLP) may improve outcomes and reduce postoperative stenosis rates. Previous studies have implicated exposed bone as a significant cause of restenosis. Tran et al. [6] reported a mean stenosis rate of 33% in 77 patients undergoing EMLP while 28.6% demonstrated a stenosis greater than 60%. Additional findings by this group suggested that the significant restenosis rate is exacerbated in the setting of increased intraoperative bone exposure [7]. The authors thereby advocated the placement of free mucosal grafts over the osteotomy site, particularly in the setting of large areas of exposed bone. The advantages of vascularized reconstruction of the anterior cranial fossa may be applied in the setting of EMLP, suggesting that coverage of the exposed bone with a pedicled mucosa flap may serve to further reduce the postoperative restenosis rate.

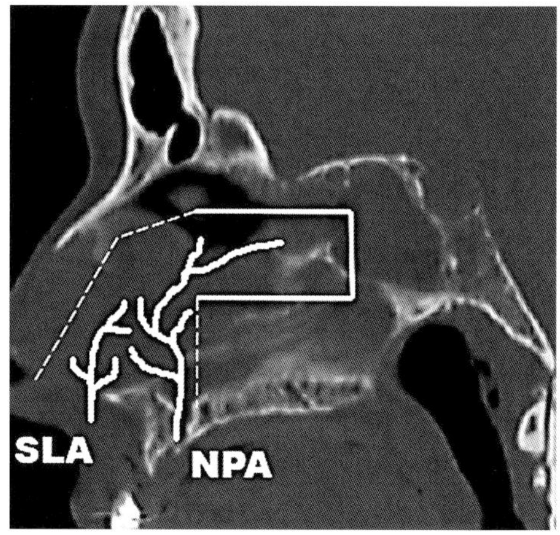

Fig. 1. Outline of the BAS flap superimposed on a sagittal CT image (solid lines). The dotted lines represent the back cuts required to roll the flap onto the nasal floor to permit protection during anterior skull base surgery. The origin of the vascular supply from the SLA and the NPA are illustrated.

Bipedicled Anterior Septal Flap

The need for vascularized tissue for reliable coverage of extremely anterior defects coupled with the limitations of posteriorly pedicled flaps implies the need for an equivalent anteriorly pedicled flap. Unfortunately, in contradistinction to the large bore SPA arborizations, the anterior nose is subserved by a vascular arcade of many small caliber vessels, none of which are substantial enough to support a large mucosal flap on their own. Thus, in order to create a vascular pedicle which is both anatomically predictable and capable of supporting a large area of mucosa the incorporation of two independent arterial inputs may be employed. Both the septal branch of the superior labial artery (SLA) and the nasopalatine artery (NPA) supply the septum along its inferior border and can be utilized to create a bipedicled anterior septal (BAS) flap (fig. 1). The nomenclature surrounding the NPA in inconstant throughout the literature; however, it represents the anastomosis between the sphenopalatine and greater palatine arteries which is transmitted through the incisive canal along with the nasopalatine nerve. The recent elucidation of Kiesselbach's plexus as an anastomotic network rather than a series of end arteries [8] suggests that a bipedicled arterial input would be capable of supporting a similar vascular bed as the posterior septal artery through retrograde filling following flap harvest.

An important criteria for the use of such a flap is that it must be able to be harvested prior to the extirpative portion of the surgery and stored in a protected position during the course of the procedure. Thus through the use of two judicious back cuts (depicted as dotted lines in fig. 1), the BAS flap and its associated pedicles may be rolled onto the nasal floor allowing for unimpeded endoscopic access to the entire anterior skull base.

In 2011, Bleier et al. [9] performed radioanatomic measurements in 9 patients as well as studies in 4 cadaveric heads to examine whether the BAS flap could be

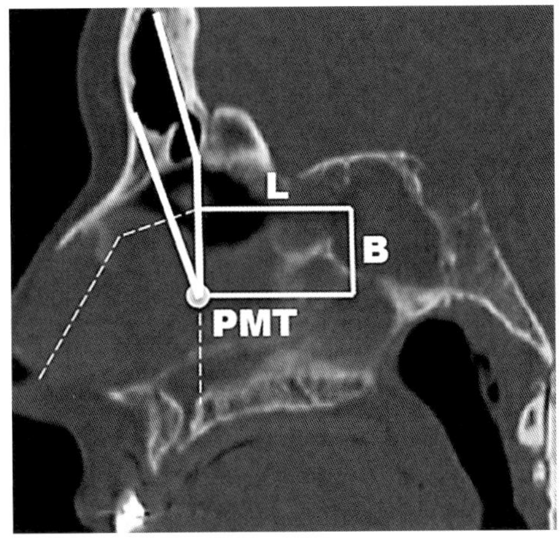

Fig. 2. Flap measurements depicted with the length (L) to base (B) relationship. Following elevation, the flap is rotated along the PMT prior to inset. The path length to the frontal beak and posterior frontal table are denoted by solid white lines.

employed as a feasible method for the reconstruction of anteriorly located skull base defects. The mean flap length (L) and base (B) were found to be 39.64 ± 4.26 and 22.81 ± 4.31 mm, respectively, and the anterior-inferior most point of the flap was designated as the point of maximal tension (PMT) (fig. 2). The BAS flap was found to be able to cover 100.31 ± 21.34% of the distance from the PMT to the top of the frontal beak, and 50.74 ± 32.54% of the posterior frontal table. With respect to the reconstruction of the frontal beak, the cadaveric data confirmed the radioanatomic measurements indicating that the BAS flap is capable of reaching the superior extent of the beak following EMLP. Placement of the flap in this region required two independent off-axis 90° rotations which led to a diminution of up to 20% of the total flap length. Despite this loss, the flap was still found to be capable of reliably covering the entire frontal beak. While insetting of the thick mucoperichondrial flap resulted in an immediate loss of neo-ostial area, the addition of vascularized tissue and consequent disruption of any cicatrical scar formation was hypothesized to be able reduce subsequent ostial stenosis.

The findings of this study confirmed that the BAS flap was capable of providing reliable coverage of defects involving the nasal beak and 50% of the height of the posterior table thereby proving a novel method for reconstruction of both postsurgical and post-traumatic posterior table injuries including those with an associated CSF leak.

The BAS flap is also advantageous in patients in which the status of the posterior septal artery may be unknown, such as following wide sphenoidotomy during functional endoscopic sinus surgery (FESS). The arterial inputs of the BAS flap are rarely encountered in routine sinus surgery thereby providing a more reliable blood supply in the post-FESS patient. The fact that the mean length to base ratio is 3.29 also suggests that the distal end of the flap may survive based on the submucosal plexus even if one of the pedicles is inadvertently injured. Additionally, raising an

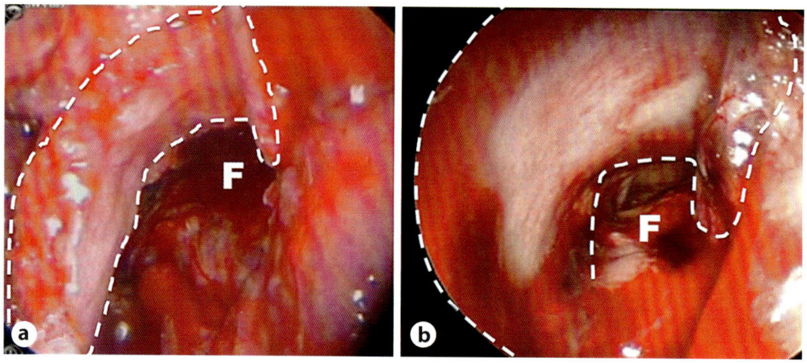

Fig. 3. **a** Intraoperative view of the BAS flap (dotted line) inset following EMLP demonstrating coverage of all exposed bone and a moderate loss of neoostial area due the flap thickness. **b** Same patient on postoperative day 6 demonstrating viability of the flap (dotted line) and a lack of any significant crusting around the drill out site.

adequate HBF often requires harvesting of septal mucosa anterior to the caudal septal margin resulting in prolonged, symptomatic crusting and the potential for internal valve obstruction. In contrast, the mucosa of the BAS flap is harvested from the posterosuperior septum resulting in a donor site which may be less symptomatic. When back cuts are required, the anterior septal mucosa is replaced once the flap is inset yielding a single incision which may closed and allowed to heal by primary intention.

The flap described in the Bleier et al. [9] study is based on a rectangular design which provides a compromise between flap length and width (i.e. base). However, the flap may be further tailored to the specific anticipated defect in a variety of ways. The distal reach may be extended by moving the PMT superiorly; however, this will necessarily sacrifice the flap width. Additionally, the total length may be extended by carrying the flap around the vomer to the contralateral posterior septum or potentially extending it across the arch of the choana towards the sphenopalatine foramen. The latter two modifications may place the viability of the distal blood supply at risk and should be attempted with caution (online suppl. video 1).

OSM

Anterior Lateral Nasal Wall Flaps

The benefits of anteriorly pedicled flaps have prompted investigations into additional flaps based on the lateral nasal blood supply. Gil and Margalir [10], in 2012, described an anterior inferior turbinate flap (AITF) for reconstruction of anterior skull base defects. This flap exploits the anterior ethmoid contributions to the inferior turbinate artery yielding a mucoperichondrial flap up to 5 cm in length and 2 cm in width. The entire length of the inferior turbinate is used, and the pedicle is based anteriorly. The authors note that this flap may be used for reconstruction of the cribiform plate, a portion of the posterior table of the frontal sinus, and the ethmoid roof. Of the 7 patients that were reconstructed with the AITF (either alone or in conjunction with a nasoseptal flap or a posteriorly based inferior turbinate flap), 100% of the flaps were viable at 8 weeks without CSF leaks or complications.

In 2012, Hadad et al. [11] described a similar anteriorly based inferior turbinate flap based on the branches of the facial, angular, and anterior ethmoidal arteries (HB2 flap). In contrast to the Gil flap, they extended the medial aspect of the flap onto the nasal floor while sparing the mucosa of Hasner's valve providing for a larger and wider flap. The authors reported the ability to reconstruct from the crista galli to the tuberculum sella in an anterior to posterior dimension and from orbit to orbit laterally. When applied to three skull base defects, no CSF leaks were reported; however, the authors did note a single episode of large volume epistaxis 10 days after surgery from the ipsilateral posterior nasal artery.

Conclusions

The increasing use of endonasal approaches for resection of intradural tumors has resulted in larger defects that must be repaired to ensure separation of the cranial vault from the nasal cavity. The workhorse of anterior skull base reconstruction remains the nasoseptal or HBF. Despite its success, the nasoseptal flap is limited in its ability to reach extremely anterior defects including those involving the frontal break, posterior frontal table, and anterior cribiform plate. Alternative approaches utilizing anteriorly pedicled flaps have been described which exploit the vascular supply to the anterior septum and lateral nasal wall. The diminutive nature of the anterior septal blood supply has led to the elaboration of a BAS flap which is capable of reliable reconstruction of both the frontal beak as well as posterior frontal table defects [9]. Similarly, Gil and Margalir [10] and Hadad et al. [11] both described anteriorly based inferior turbinate flaps pedicled largely on anterior ethmoid arborizations which have been used successfully to reconstruct anterior defects. In light of the limitations of the nasoseptal flap, the addition of anteriorly pedicled flaps to the clinical armamentarium offers the opportunity to provide vascularized mucosal coverage of virtually any region of the skull base which can be reached endoscopically.

References

1 Hadad G, Bassagasteguy L, Carrau RL, Mataza JC, Kassam A, Snyderman CH, Mintz A: A novel reconstructive technique after endoscopic expanded endonasal approaches: vascular pedicle nasoseptal flap. Laryngoscope 2006;116:1882–1886.

2 Patel MR, Stadler ME, Snyderman CH, Carrau RL, Kassam AB, Germanwala AV, Gardner P, Zanation AM: How to choose? Endoscopic skull base reconstructive options and limitations. Skull Base 2010; 20:397–404.

3 Prevedello DM, Barges-Coll J, Fernandez-Miranda JC, et al: Middle turbinate flap for skull base reconstruction: cadaveric feasibility study. Laryngoscope 2009;119:2094–2098.

4 Fortes FS, Carrau RL, Snyderman CH, et al: The posterior pedicle inferior turbinate flap: a new vascularized flap for skull base reconstruction. Laryngoscope 2007;117:1329–1332.

5 Virgin F, Barañano CF, Riley K, Woodworth BA: Frontal sinus skull base defect repair using the pedicled nasoseptal flap. Otolaryngol Head Neck Surg 2011;145:338–340.

6 Tran KN, Beule AG, Singal D, et al: Frontal ostium restenosis after the endoscopic modified Lothrop procedure. Laryngoscope 2007;117:1457–1462.

7 Seiberling K, Floreani S, Robinson S, et al: Endoscopic management of frontal sinus osteomas revisited. Am J Rhinol Allergy 2009;23:331–336.

8 Chiu T, Dunn JS: An anatomical study of the arteries of the anterior nasal septum. Otolaryngol Head Neck Surg 2006;134:33–36.

9 Bleier BS, Curry WT, Wang EW, Schlosser RJ: The bipedicled anterior septal flap: a radioanatomic and cadaveric study. Laryngoscope 2011;121:1367–1371.

10 Gil Z, Margalir N: Anteriorly based inferior turbinate flap for endoscpic skull base reconstruction. Otolaryngol Head Neck Surg 2012;146:842–847.

11 Hadad G, Rivera-Serrano CM, Bassagaisteguy LH, Carrau RL, Fernandez-Miranda J, Prevedello DM, Kassam AB: Anterior pedicle lateral nasal wall flap: a novel technique for the reconstruction of anterior skull base defects. Laryngoscope 2011;121: 1606–1610.

Benjamin S. Bleier, MD
Department of Otology and Laryngology, Massachusetts Eye and Ear Infirmary
Harvard Medical School
243 Charles St., Boston MA 02114 (USA)
E-Mail bleierb@gmail.com

OSM

Bleier BS (ed): Comprehensive Techniques in CSF Leak Repair and Skull Base Reconstruction.
Adv Otorhinolaryngol. Basel, Karger, 2013, vol 74, pp 71–80 (DOI: 10.1159/000342282)

Pedicled Extranasal Flaps in Skull Base Reconstruction

Grace G. Kim · Anna X. Hang · Candace A. Mitchell · Adam M. Zanation

Department of Otolaryngology/Head and Neck Surgery, University of North Carolina Hospitals, Chapel Hill, N.C., USA

Abstract

Cerebrospinal fluid (CSF) leaks most commonly arise during or after skull base surgery, although they occasionally present spontaneously. Recent advances in the repair of CSF leaks have enabled endoscopic endonasal surgery to become the preferred option for management of skull base pathology. Small defects (<1 cm) can be repaired by multilayered free grafts. For large defects (>3 cm), pedicled vascular flaps are the repair method of choice, resulting in much lower rates of postoperative CSF leaks. The pedicled nasoseptal flap (NSF) constitutes the primary reconstructive option for the vast majority of skull base defects. It has a large area of potential coverage and high rates of success. However, preoperative planning is required to avoid sacrificing the NSF during resection. In cases where the NSF is unavailable, often due to tumor involvement of the septum or previous resection removing or compromising the flap, other flaps may be considered. These flaps include intranasal options – inferior turbinate or middle turbinate flaps – as well as regional pedicled flaps: pericranial flap, temporoparietal fascial flap, or palatal flap. More recently, novel alternatives such as the pedicled facial buccinator flap and the pedicled occipital galeopericranial flap have been added to the arsenal of options for skull base reconstruction. Characteristics of and appropriate uses for each flap are described.

In the last 10 years, anterior skull base reconstruction to repair cerebral spinal fluid (CSF) leaks has evolved tremendously. The primary goal of skull base reconstruction is to obtain a watertight closure, separating the cranium and the sinonasal cavity as a barrier against intracranial infection. Other goals include elimination of dead space, preservation of neurovascular and ocular function, and promotion of wound healing.

Early success with endoscopic reconstructive techniques to repair spontaneous CSF leaks caused by accidental or iatrogenic trauma has led to acceptance of endoscopic transnasal transcranial surgery as the preferred option for managing benign and malignant diseases. Small (<1 cm) defects in the skull base, commonly found during CSF fistula closure, are reliably repaired (success rate greater than 90%) using multilayered free grafts [1]. In such small defects, merely establishing a tissue barrier is sufficient. However, larger (>3 cm) defects such as those commonly caused by tumors (e.g. meningioma) are more difficult to repair. The repair of large defects using free grafts has resulted in rates of postoperative CSF leaks up to 40% [2], which varies by tumor location.

Large defects present additional challenges including wide resection of the dura, additional dissection of the arachnoid, and greater exposure to high-flow CSF. With advancements in reconstructive techniques using local and regional vascularized flaps, the rate of postoperative CSF leak has been reduced to less than 5% [3]. In this chapter, we discuss the use of pedicled extranasal flaps that may improve management of large skull base defects.

Clinical Presentation

A patient's history is most suggestive of postoperative CSF leak, which may occur spontaneously after trauma or, more commonly, within a week following skull base surgery. Patients may present with the complaint of experiencing metallic or salty taste and symptoms including rhinorrhea or clear postnasal drip. On physical exam, findings suggestive of CSF leak include a positive reservoir sign (clear rhinorrhea induced by the patient leaning forward) or rhinorrhea elicited by the Valsalva maneuver. Sinonasal endoscopy can help localize the CSF leak. Assay of rhinorrhea fluid for beta-2-transferrin, a protein found in CSF, may be used as a confirmatory test [4].

It is less common to experience a postoperative CSF leak beyond a week from surgery. Intermediate timed CSF leaks (2–6 weeks postoperatively) present with intermittent low flow leaks from a small dural opening. Late CSF leaks (more than 6 weeks) are rare and usually seen with hard nose blowing, radionecrosis, and prior surgery [5].

High Risk Factors for Postoperative CSF Leak

Risk for postoperative CSF leak is often dependent on size and location of defect, intraoperative dural violation, and the type of reconstruction employed. Large defects repaired with free grafts are associated with unacceptably high rates of CSF leak, >30%, while small defects have much lower risk of developing postoperative CSF leak regardless of method of repair [6]. In terms of location, anterior skull base defects are much more likely to leak than clival defects [5]. The presence of an intraoperative high-flow CSF leak is the most reliable predictor of developing a postoperative leak

[7]. Thus, pathology that requires dissection into the arachnoid cisterns or ventricles is also associated with higher rates of postoperative CSF leaks. However, with the advent of vascularized flap reconstruction, the rate of postoperative leaks has been dramatically reduced to less than 5% [7, 8]. Patient factors that increase the risk of a post-operative CSF leak include [5]:

1. Obesity – associated with increased ventricular pressure.
2. Cushing's disease – increased circulating cortisol levels can lead to poor tissue healing.
3. History of prior surgery or radiation – limits options for vascularized tissue reconstruction due to impaired tissue healing.

Management of Skull Base Defects

The endoscopic expanded endonasal approach (EEA) is now widely used to treat a full range of extradural and intradural pathologies. As the skull base defects created by EEAs became more complex, the options for reliable reconstruction expanded as well. Materials for reconstruction now include avascular grafts, pedicled nasoseptal flaps (NSF), turbinate flaps, and endoscopic regional flaps from extranasal sources. The choice of appropriate reconstructive options should be guided by the location and size of the defect, type of intraoperative CSF leak after resection, and history of radiation or sinonasal surgery.

Prior to the routine use of vascularized tissue flaps for skull base reconstruction, free grafts of biologic or synthetic material were used as the primary reconstructive technique. The key to a successful repair was use of a multilayer approach. The repair begins with a subdural inlay graft of collagen matrix (Duragen) to obliterate the intradural dead space, followed by an inlay graft of acellular dermis (AlloDerm) in the epidural space. In many cases of EEA skull base defects, especially at the sphenoid or clivus, the bony ledges are limited and cannot support an inlay graft. In these cases, an extracranial onlay graft may be used. The grafts are then bolstered intranasally with absorbable packing, synthetic or biologic glue, and a nasal Foley catheter balloon or expandable sponge packing. This technique can be modified for moderately sized defects by using an onlay free mucosal graft instead of the Alloderm graft. Abdominal fat can also be added to the multilayer reconstruction as an additional bolster or biologic dressing or to obliterate dead space. For small defects (<1 cm), the multilayer approach has a success rate of greater than 90% [6]. For defects greater than 3 cm, reconstruction with multilayered free tissue grafts results in unacceptably high rates of postoperative CSF leaks of 20–30%, and therefore not recommended [3]. Reconstruction of the bony defect following skull base surgery is controversial. Advocates of reconstruction cite the need to prevent herniation of cranial contents or pulsatile exophthalmos. However, much literature has shown that the bony defect presents no functional or aesthetic consequence [9]. If structural support is needed, titanium mesh, calvarial bone, or split rib grafts have been used in the past. However, introduction of free bone grafts or synthetic material can lead to poor tissue healing [10].

Reconstruction with a vascular pedicled flap is preferred for large defects resulting from wide dural resection with associated extra arachnoid dissection, high flow CSF leaks, and a nonvascularized reconstructive bed. Of the local and regional vascularized flaps, the NSF has increasingly become the workhorse of endonasal skull base reconstruction. The NSF, a vascular flap composed of mucoperiosteum and mucoperichondrium derived from the nasal septum and pedicled on the posterior nasoseptal artery [5], are discussed in detail in the chapter by Snyderman and Pinheiro-Neto [this vol.]. The advantages of the NSF include endoscopic graft harvest avoiding a second incision and the ability to cover a most of anterior skull base defects based on radioanatomic studies [12]. The major disadvantage is that the use of the NSF must be anticipated preoperatively: without this planning, the dissection might compromise vascular supply to the flap during sphenoidotomy or posteror septectomy. This flap may not be available secondary to sinonasal tumor involvement or if prior surgery has used the flap or compromised the pedicle. Additionally, in the pediatric population (<10 years of age), the NSF area is often significantly smaller than the age-corresponding skull base defect because septal growth is not completely developed until puberty [13].

For patients in whom an NSF is inadequate or unavailable, an inferior turbinate (ITF) or middle turbinate flap (MTF) can be considered. The ITF and MTF are posteriorly pedicled from the inferior turbinate artery and middle turbinate artery respectively, both of which are branches off the posterior lateral nasal artery, a terminal branch of the sphenopalatine artery [5]. The entire medial surface of the inferior turbinate should be harvested to maximize coverage. The harvest can even be extended to include the lateral mucoperiosteum for a wider flap. The ITF can cover approximately 60% of the anterior cranial fossa and bilateral flaps can be used for larger defects [5]. Although coverage may be incomplete in some cases, reconstruction can still be successful if augmented with free grafts owing to the vascularity of the flap. Compared to the NSF, the ITF is shorter in length and has a limited arc of rotation. Due to these limitations, the ITF is best suited for small (<1 cm) posterior defects of the sella, parasellar and midclival areas [5]. Silicone nasal splints are typically left in place for 10 to 21 days to protect the denuded lateral nasal wall. The MFT is another posteriorly pedicled local flap that can be used small anterior skull base defects. Elevation of the MFT is technically challenging, especially in the setting of anatomic variations, and only produces a small flap of thin layer of mucosa [7]. However, due to the more superior position of its pedicle, the MTF is better suited than the ITF for defects of the planum sphenoidale, sella, and fovea ethmoidalis [5]. In general, the MTF is used sparingly as a secondary option if a NSF is not available.

As endoscopic resections for skull base pathology become more complex, the resultant defects require more difficult and extensive reconstructions. In such cases, use of regional tissue flaps, distal pedicled flaps, or even free tissue transfer may be indicated. Extranasal flaps are ideal for endonasal reconstruction because they are not involved in the primary cancer site [8]. Details of regional extranasal flaps such as the frontal or occipital pericranial flap, temporoparietal fascial flap, palatal flap, or facial buccinator flap are described later in the chapter. Distal pedicled fasciocutaneous, muscle, or myocutaneous flaps include pectoralis major, trapezius, latissimus dorsi, or sternocleidomastoid flaps. However, their vascular attachments originate below the clavicle, which limits their reach to the skull base. The use of a vascularized free flap allows for great flexibility in flap content and design, a single stage reconstruction, and the ability to introduce a large quantity of vascularized tissue to eliminate dead space. Pedicled flaps are often limited by length and arc of rotation allowed by the pedicle and the fact that the distal, most tenuous portion of the flap is often positioned over the most crucial portion of the defect. Both of these issues are eliminated with the use of a free flap. Free flaps also have the potential to provide enough tissue for coverage of defects that extend to the middle cranial fossa [11]. Donor sites for free flaps include the radial forearm, rectus abdominis, latissimus dorsi, serratus, scapula or gracilis [10]. The recipient vessels are often the superficial temporal vessels. Detailed discussion of free vascularized flap reconstruction of skull base defects is covered in a separate chapter. A summary of local and regional vascular flaps options can be found in table 1.

Pedicled Extranasal Flap Options

Although a pedicled NSF is the reconstructive option of choice for most endoscopic skull base defects, other vascularized flaps may be necessary in some cases to achieve optimal outcomes. This situation arises almost exclusively in cases where the NSF is unavailable, due either to previous surgical resection or to involvement of the nasal septum by tumor or diseased tissue. When the NSF is not a reconstructive option, a regional pedicled flap becomes the preferred reconstructive option. Such flaps are most commonly harvested from the pericranium (pericranial flap, PCF), temporoparietal fascia (temporoparietal fascial flap, TPFF), or from the palatal mucosa (palatal flap, PF). Additionally, novel options such as the pedicled facial buccinator flap (FAB) and the pedicled occipital galeopericranial flap (OGP) have recently been described. All of these alternative vascularized flaps are preferred to avascular or artificial repair options because, as with the NSF, the rate of CSF leak is lower with vascularized repair.

Pericranial Flap (online suppl. video 1)

The PCF is the most versatile alternative to the NSF and has the potential for the largest area of coverage. Its pedicle is based upon the supraorbital and supratrochlear arteries, and it was commonly used for skull base reconstruction in the era prior to

Table 1. Intranasal and regional vascular flaps available for skull base reconstruction

Location	Vascular tissue flap	Pedicle	Advantages/uses	Disadvantages/limitations
Intranasal	NSF	posterior septal artery (from SPA)	ideal for all skull base reconstructions	size limitation in pediatric patients
	ITF	inferior turbinate artery*	good for small clival defects	cannot reach ACF or sella crusting over denuded turbinate
	MTF	middle turbinate artery*	good for small ACF or transsphenoidal defects	small in size thin mucosa difficult to elevate
Regional	PCF	supraorbital and supratrochlear arteries	hearty flap with versatile dimensions extends from ACF to sella	does not extend to posterior skull base requires external incision
	TPFF	superficial temporal artery	good for clival or parasellar defects usually from nonirradiated field	90° pedicle rotation limits reconstruction of ACF requires external incision
	PF	greater palatine artery	theoretically can reach all areas of skull base long (3 cm) pedicle	difficult to dissect pedicle donor site morbidity unproven clinically
	FAB	facial artery	useful for anterior skull base no facial incision	potential paresthesia or persistent epiphora potential introduction of oral flora to sterile field unproven clinically
	OGP	occipital artery	good for posterior lesions potential for very large area of coverage long (8+ cm) pedicle usually from nonirradiated field	potential for carotid injury with proximal dissection of pedicle unproven clinically

* Terminal branch of posterior lateral nasal artery of the sphenopalatine artery.
SPA = Sphenopalatine artery; NSF = nasoseptal flap; ITF = inferior turbinate flap; MTF = middle turbinate flap, PCF = <lap; TPFF = temporoparietal fascia flap; PF = palatal flap; FAB = <; OGP = occipital galeopericranial flap; ACF = anterior cranial fossa.

the widespread use of endoscopic resection [5]. At first, it was considered to have little utility for endoscopic skull base surgery. By transposing the flap through a small nasionectomy into the endoscopic field, this flap became a useful alternative to the NSF [5]. In its current form, it is best suited for defects of the anterior skull base: it can cover an extensive area and its pedicle allows extension from the anterior cranial fossa as far as the sella [7]. However, it is unable to reach posterior cranial base defects. Historically, this flap has entailed large external incisions and suboptimal cosmesis; however, endoscopic harvesting of this flap [14] has allowed for minimal adverse effect on cosmesis as well as better visualization during flap rotation to avoid torsion of the flap pedicle [7].

Results of endoscopically harvested PCF are quite promising: Patel et al. [15] described a series of 10 reconstructions by their senior authors. In their series, there were no postoperative CSF leaks, no V1 paresthesia, no facial nerve dysfunction, and no significant cosmetic deformities. Furthermore, frontal sinus patency was preserved in all cases and all flaps covered 100% of the cranial defect. The flap survived well after radiation treatment; there were no cases of late flap complications (>3 months) in their series.

Special considerations while using the PCF include ensuring that the flap passes through the glabellar incision without twisting the pedicle and compromising the vascular supply of the flap. When extending the flap for larger skull base defects, it is critical not to extend the PCF past midline in order not to compromise contralateral structures which could become necessary for revision surgery [15]. Additionally, it is important to note that this reconstructive option becomes more challenging in older patients due to difficulty dissecting within the subperiosteal plane; however, authors have documented successful flap harvesting in patients up to 80 years old [15]. Finally, when using this flap, postoperative treatment must avoid use of a compressive cranial dressing: this intervention may compromise the blood flow through the PCF pedicle [15].

Temporoparietal Fascial Flap (online suppl. video 2)

For defects of the sella, parasella and clivus, the TPFF provides a good alternative reconstructive option [6, 7]. The superficial temporal artery (STA), a terminal branch of the external carotid artery, comprises blood supply for the flap. The flap is harvested after endoscopic dissection is complete so that the exact dimensions needed for reconstruction are known prior to harvesting. At its largest, the dimensions of the fan-shaped flap can reach 17 × 14 cm [5]. Thus, the TPFF can comprise a large, durable reconstructive option that is especially advantageous for patients who have undergone sinonasal or skull base radiotherapy, as this vascularized tissue lies beyond the irradiated field [5].

The flap is limited by the arc of rotation of its pedicle, which must rotate 90 degrees as it is passed through the pterygopalatine fossa. It therefore is poorly suited for reconstruction of defects of the anterior cranial fossa [7]. Additionally, this flap does require an external incision, leading to a surgical scar and accompanying risk of alopecia [5]. While harvesting the flap, careful attention is necessary to avoid injury to the pedicle, as the STA lies directly below the skin near the hemicoronal incision created. Care must also be taken during flap harvesting not to injure the facial nerve, which courses just below the TPFF after crossing the superficial surface of the zygomatic arch [5].

Palatal Flap

An additional reconstructive option that has recently been described is the PF. For this flap, the vascularized mucoperiosteal tissue of the hard palate is passed through the greater palatine foramen into the nasal cavity to reach defects of the skull base [5]. Theoretically, this flap could prove incredibly useful: its vascular supply derives from the greater palatine artery, providing a 3-cm pedicle that could potentially reach any area of the skull base [7]. However, to date the technique has mainly been described in cadaveric studies [7]. This lack of clinical experience and the potential donor site morbidity incurred by removal of the hard palate mucosa have, to date, relegated this flap to use as a reconstructive option of last resort [5].

Pedicled Facial Buccinator Flap

The pedicled FAB was recently described by Rivera-Serrano et al. [16]. This flap, as its name suggests, is pedicled upon the facial artery after it branches off the external carotid artery. In cadaveric studies, the FAB had a length of 7–8 cm from pedicle to tip and was able to reach the anterior skull base and planum sphenoidale [16]. Its size is limited by anatomical considerations in the area of harvest. The parotid duct orifice creates a superior limit, although by extending the flap in an 'L' or 'Boot' configuration around this landmark, the potential size is increased [16].

Two options for harvesting the FAB exist: solely muscular (facial artery musculo-buccinator, FAMB) or combined myomucosal (facial artery musculo-mucosal-buccinator, FAMMB) flaps. When mucosa is used, it becomes necessary to rotate the flap 180 degrees in order to place the mucosal surface within the nasal cavity [16]. Thus, the authors suggest that the FAMB flap is preferable for skull base reconstruction due to the potential for venous obstruction secondary to this rotation [16]. For either option, transposition of the FAB flap takes place through a maxillary window.

Because the FAB flap has only been described in cadaveric studies to date, the risk profile is largely theoretical. One potential complication would be the risk of dental or facial paresthesia [16]. Dissection of this flap does traverse territory close to facial nerve branches; however, because dissection should take place in the plane directly superficial to the vascular pedicle, facial nerve injury should not be expected in these patients [16]. Other potential complications include persistent epiphora and introduction of oral flora into the surgical field. The authors propose to mitigate these risks by installing Crawford silicone tubes to address epiphora and by instituting a protocol of chlorohexedine gargle 4 hours prior to surgery to decrease bacterial counts [16].

Occipital Galeopericranial Flap

Another alternative recently touted by Rivera-Serrano e al. [17] for its potential in skull base reconstruction is the OGP. This flap is an excellent choice for large skull base lesions, as flaps as large as 44 cm^2 (11×4 cm) were harvested in their cadaveric studies [17]. This expansive flap is pedicled upon the occipital artery, which supplies the majority of the posterior scalp. Use of this large artery is advantageous, providing a long pedicle and thus enhancing the versatility of the flap [17]. Dissection of the pedicle is facilitated by detachment of the overlying neck musculature, including the sternocleidomastoid [18]. Furthermore, the anatomy of the blood supply to the OGP shows a high rate of consistency between individuals, facilitating a reliably successful dissection [18]. Transposi tion takes place through a tunnel along the inferomedial surface of the medial pterygoid muscle. Use of the flap therefore requires a wide endoscopic maxillary antrostomy and posterior maxillectomy as well as removal of the inferior aspect of the pterygoid plates [18]. Finally, the distance of the flap from the skull base itself makes the OGP an excellent choice in patients who have previously undergone surgery or radiotherapy [18].

The proposed use for the OGP is largely relegated to lesions of the posterior skull base due to the success of the PCF in reconstructing large lesions of the anterior skull base, as discussed above. In addition, more ventral positioning of the OGP increases the risk of compression of bulkier, more distal portions of the flap, and may pose a threat of venous congestion and flap compromise [18]. Of note, harvesting the OGP does require transection or detachment of several neck muscles, including the sternocleidomastoid and trapezius. However, Dogliotti and Bennum [18] described favorable outcomes with reattachment in their use of this flap for mandibular reconstruction, noting no residual neck muscle weakness. Dissection of the pedicle to its proximal extent might also pose the threat of carotid injury; however, ample pedicle length is available without such extensive dissection [18].

Conclusion

Anterior skull base defects can be successfully and reliably repaired via a purely endoscopic endonasal approach. This chapter provides a comprehensive overview of extranasal reconstructive options for skull base repair. With advances of new reconstruction techniques and use of vascularized flaps, postoperative CSF leak rates have been reduced to less than 5%. Pedicled extranasal flaps allow a better reconstructive option for larger and more complex lesions particularly when vascularized mucosal flaps are unavailable or inadequate.

Acknowledgements

Supported by a grant for the National Institute on Deafness and other Communicative Disorders, T32DC005360 (G.G.K.).

Disclosure Statement

None.

References

1 Hegazy HM, Carrau RL, Snyderman CH, Kassam A, Zweig J: Transnasal endoscopic repair of cerebrospinal fluid rhinorrhea: a meta-analysis. Laryngoscope 2000;110:1166–1172.

2 Gardner PA, Kassam AB, Thomas A, Snyderman CH, Carrau RL, Mintz AH, Prevedello DM: Endoscopic endonasal resection of anterior cranial base meningiomas. Neurosurgery 2008;63:36–52.

3 Hadad G, Bassagasteguy L, Carrau RL, Mataza JC, Kassam A, Snyderman CH, Mintz A: A novel reconstructive techinique after endoscopic expanded endonasal approaches: vascular pedicle nasoseptal flap. Laryngoscope 2006;116:1882–1886.

4 Skedros DG, Cass SP, Hirsch BE, Kelly RH: Beta-2 transferrin assay in clinical management of cerebral spinal fluid and perilymphatic fluid leaks. J Otolaryngol 1993;22:341–344.

5 Zanation AM, Thorp BD, Parmar P, Harvey RJ: Reconstructive options for endoscopic skull base surgery. Otolaryngol Clin North Am 2011;44: 1201–1222.

6 Harvey RJ, Parmar P, Sacks R, Zanation AM: Endoscopic skull base reconstruction of large dural defects: a systematic review of published evidence. Laryngoscope 2012;122:452–459.

7 Patel MR, Stadler ME, Snyderman CH, Carrau RL, Kassam AB, Germanwala AV, Gardner P, Zanation AM: How to choose? Endoscopic skull base reconstructive options and limitations. Skull Base 2010; 20:397–404.

8 Zanation AM, Carrau RL, Snyderman CH, Germanwala AV, Gardner PA, Prevedello DM, Kassam AB: Nasoseptal flap reconstruction of high flow intraoperative cerebral spinal fluid leaks during endoscopic skull base surgery. Am J Rhinol Allergy 2009;23:518–521.

9 Chang DW, Langstein HN, Gupta A, De Monte F, Do KA, Wang X, Robb G: Reconstructive management of cranial base defects after tumor ablation. Plast Reconstr Surg 2001;107:1346–1355.

10 Gagliardi F, Boari N, Mortini P: Reconstruction techniques in skull base surgery. J Craniofac Surg 2011;22:1015–1020.

11 Georgantopoulou A, Hodgkinson PD, Gerber CJ: Cranial-base surgery: a reconstructive algorithm. Br J Plast Surg 2003;56:10–13.

12 Pinheiro-Neto CD, Prevedello DM, Carrau RL, Snyderman CH, Mintz A, Gardner P, Kassam A: Improving the design of the pedicled nasoseptal flap for skull base reconstruction: a radioanatomic study. Laryngoscope 2007;117:1560–1569.

13 Shah RN, Surowitz JB, Patel MR, Huang BY, Snyderman CH, Carrau RL, Kassam AB, Germanwala AV, Zanation AM: Endoscopic pedicled nasoseptal flap reconstruction for pediatric skull base defects. Laryngoscope 2009;119:1067–1075.

14 Zanation AM, Snyderman CH, Carrau RL, Kassam AB, Gardner PA, Prevedello DM: Minimally invasive endoscopic pericranial flap: a new method for endonasal skull base reconstruction. Laryngoscope 2009;119:13–18.

15 Patel MR, Shah RN, Snyderman CH, Carrau RL, Germanwala AV, Kassam AB, Zanation AM: Pericranial flap for endoscopic anterior skull-base reconstruction: clinical outcomes and radioanatomic analysis of preoperative planning. Neurosurgery 2010;66:506–512.

16 Rivera-Serrano CM, Oliver CL, Sok J, et al: Pedicled facial buccinator (FAB) flap: a new flap for reconstruction of skull base defects. Laryngoscope 2010; 120:1922–1930.

17 Rivera-Serrano CM, Snyderman CH, Carrau RL, et al: Transparapharyngeal and transpterygoid transposition of a pedicled occipital galeopericranial flap: a new flap for skull base reconstruction. Laryngoscope 2011;121:914–922.

18 Dogliotti PL, Bennun RD: Occipitoparietal bone flap for mandibular reconstruction. Journal Craniofac Surg 1995;6:249–254.

Adam M. Zanation, MD
Department of Otolaryngology/Head and Neck Surgery, University of North Carolina
170 Manning Drive, CB 7070, Physician's Office Building, Room G-190
Chapel Hill, NC 27599–7070 (USA)
E-Mail adam_zanation@med.unc.edu

Bleier BS (ed): Comprehensive Techniques in CSF Leak Repair and Skull Base Reconstruction.
Adv Otorhinolaryngol. Basel, Karger, 2013, vol 74, pp 81–91 (DOI: 10.1159/000342283)

Microvascular Free Flaps in Skull Base Reconstruction

Marc W. Herr · Derrick T. Lin

Massachusetts Eye and Ear Infirmary, Harvard Medical School, Boston, Mass., USA

Abstract

The anatomical challenges of skull base surgery are well known. Furthermore, ablative and traumatic defects in this region produce complex reconstructive problems with a high risk of significant postoperative morbidity and mortality. Over the past two decades, microvascular free tissue reconstruction following open resection has been shown to improve outcomes and reduce complication rates when compared to the traditional use of pedicled flaps. The increasing use of free tissue transfer has been further strengthened by improved technical expertise and high flap success rates. Since the size and type of free tissue to be utilized must be individualized to each defect, the accomplished reconstructive surgeon should be extremely versatile and, by extension, facile with a several types of free flaps. Thus, four of the most commonly used flaps – the rectus abdominis, radial forearm, latissimus dorsi and anterolateral thigh flaps – are discussed.

The anatomy of the skull base is extremely complex and is comprised of a multitude of vital structures, including brain, dura, the orbit and its contents, the palate, the mandible, and the overlying skin. The difficult anatomical challenges produced by defects in this region are further compounded by the high risk of complications that result from skull base surgery and reconstruction. In the literature spanning over four decades, complication rates are reported to range from 25 to 65% [1–7]. A recent international review of 1,193 patients from 17 institutions categorized complication rates as follows: wound (19.8%), central nervous system (CNS)-related (16.2%), orbital (1.7%), and systemic (4.8%) [8]. These complications can be devastating resulting in CSF leak, meningitis, osteomyelitis, blindness, or delays in adjuvant treatment and producing postoperative mortality rates of up to 5.7% [1, 5, 6, 8–11]. Furthermore, even with detailed anatomic reconstruction, skull base defects often produce significant facial disfigurement. This, in turn, results in profound psychological morbidity in many patients [12]. Therefore, the goals of skull base reconstruction are: (1) to create a clear separation of the CNS from the aerodigestive tract, (2) to support the brain and orbit, (3) to re-establish the lining and spatial integrity of the nasal and oropharyngeal cavities, (4) to obliterate dead space with coverage of vital structures, and (5) to restore the three-dimensional appearance of the face and head with bone and soft tissues [1, 10, 11].

Defect Classification

Skull base defects may be classified based upon their anatomic location and loss of volume, support, and skin coverage [1]. There are two basic types – those involving the anterior cranial fossa and those involving the middle cranial fossa. The anterior cranial fossa (ACF) supports the frontal lobes of the brain intracranially and encompasses the roof of the ethmoids and orbits, as well as the cribiform plate. Thus, ACF defects result from resection of tumors involving the orbit and/or sinonasal tract. They produce a loss of orbital content, maxillary skeleton, nasal lining, bony skull base, and dura.

The middle cranial fossa (MCF) separates the temporal lobes of the brain and the infratemporal fossa. Anatomically, the region contains or borders the optic canal, the superior orbital fissure, the cavernous sinus, and the trigeminal ganglion. Defects of the MCF are often associated with tumors in the vicinity of the temporal bone. They may include large cutaneous defects and can involve the auricle, temporal bone, maxilla, and/or mandible [1, 9].

Reconstructive Options

McLean and Buncke [13] documented the first use of a free flap for skull base reconstruction when they employed omentum to cover a large scalp defect. In 1996, Neligan et al. [14] compared the outcomes of open skull base reconstruction utilizing local, pedicled, and free flaps in 90 patients over a 10-year period. They found that overall complication rates were lower in the local and free flap groups (38.8 and 33.5%, respectively) than in the pedicled flap group (75%). Furthermore, compromised healing at the skull base occurred in 36.3% of pedicled flaps with 3 patients suffering total flap loss. Conversely, there was only a 10% incidence of compromised healing in the free flap group with no flap losses [14]. Since then, the use of vascularized, free tissue has largely replaced regional muscular and musculocutaneous flaps as the preferred modality for reconstructing complex cranial base defects [1, 9, 10, 15].

There are several factors contributing to the improved outcomes seen with free tissue transfer reconstruction of the skull base. First, the size and location of skull base defects often approximates or exceeds the excursional limits of pedicled musculocutaneous flaps. Additionally, the distal tip of the pedicled flap, where the blood supply is most precarious, is usually the part of the flap that reaches the defect [9]. This places the most critical area of coverage at greatest risk for breakdown and wound exposure. On the other hand, free flaps provide well-vascualrized soft tissue to separate the intracranial contents from the upper aerodigestive tract, obliterate dead space, and cover cutaneous, mucosal, and bony defects [9].

The ideal reconstructive technique for any defect is dependent on its size and complexity. Simple and endonasal defects are amenable to repair with a vascularized mucosal, galeal-pericranial, or temporoparietal fascia (TPPF) flaps. Larger, more complex defects are best repaired with free tissue transfer, often in conjunction with adjacent pedicled flaps. This includes ablative defects greater than 40 cm^3 involving skin, the ACF, infratemporal fossa, maxilla, or lateral skull base, as well as those producing significant exposure of the paranasal sinuses and orbit or occurring in a previously radiated field [10, 16, 17].

Microvascular free tissue transfer is facilitated by an ipsilateral lymphadenectomy of levels I, II and III [1]. This provides important diagnostic staging regarding regional tumor spread and gives superior exposure to a range of possible arterial and venous recipient beds. The preferred recipient arteries include the facial, superficial temporal and external carotid arteries. The external jugular vein and facial vein serve as the primary recipient veins and are usually anastomosed to the pedicle vein in an end-to-end fashion. The internal jugular vein may also be used, in which case the anastomosis is oriented end-to-side.

In general, four free flaps can be used to reconstruct the majority of skull base defects: the rectus abdominis myocutaneous (RAM) flap, the radial forearm flap (RFF), the latissimus dorsi flap, and the anterolateral thigh (ALT) flap (table 1).

Flap Utilization and Design

Rectus Abdominis Myocutaneous Flap

With its large size, long and consistent vascular pedicle, and low donor site morbidity, the rectus abdominis myocutaneous (RAM) flap is the historic workhorse of skull base reconstruction [1, 4, 9, 11, 14, 15, 17, 18]. The blood supply of this flap is the inferior epigastric artery and vein, which enter the lateral border of the rectus muscle just below the arcuate line and halfway between the pubis and umbilicus. Dissection is performed with the patient supine. First, the cephalad portion of the flap is incised down to the level of the anterior rectus sheath. The rectus abdominis muscle is identified and an incision through the fascia is made 1 cm from the lateral edge of the muscle. The rectus is exposed from the linea alba to the linea semilunaris and dissection is carried around the anterior and lateral surfaces of the muscle to the posterior surface [1, 25]. The muscle is then dissected away from the abdominal wall in a distal to proximal direction along the posterior rectus sheath toward the inferior

Table 1. Microvascular free tissue transfer for skull base reconstruction (adapted from Pusic et al. [1])

Authors	Year	Institution	Number of patients	RAM	RFF	Lat.	ALT
Izquierdo et al. [18]	1992	Loyola	19	18		1	
Clayman et al. [19]	1995	MD Anderson	41	19	1	15	
Neligan et al. [14]	1996	Toronto Hospital	20	17	3		
Aspoas et al. [20]	1997	Queen Mary's, London	23	6	11	6	
Disa et al. [9]	1998	Memorial Sloan-Kettering	18	14		2	1
Dias et al. [4]	1999	Hospital do Cancer	45	41			
Chang et al. [17]	2001	MD Anderson	52	39	3	7	
Djalilian et al. [21]	2002	Minnesota	4	2	2		
Califano et al. [15]	2003	Memorial Sloan-Kettering	29	22	1	6	
Reza Nouraei et al. [22]	2006	Charing Cross Hospital, Liverpool	53	9	30	12	
Malata et al. [23]	2006	Addenbrooke's Hospital, Cambridge	12				8
Chiu et al. [24]	2008	Memorial Sloan-Kettering	70	68	1	1	

RAM = Rectus abdominus myocutaneous flap; RFF = radial forearm flap; Lat. = latissimus dorsi flap; ALT = anterolateral thigh flap.

epigastric pedicle [1]. The insertion of the muscle is detached and the harvest is completed.

The primary disadvantage of the RAM flap is potential bulkiness when a skin paddle is incorporated due to the width of the subcutaneous tissues of the abdominal wall. This is particularly true in women and obese patients. There is also the long-term postoperative risk of abdominal wall weakness and development of a ventral hernia, but this is minimized with proper closure of the abdominal wall defect.

Radial Forearm Flap

The radial forearm flap (RFF) is a fasciocutaneous flap composed of thin, pliable tissue and a long, consistent vascular pedicle. These features make it extremely versatile when reconstructing skull base defects with complex geometry. It is based on the radial artery with its vena comitantes and/or the cephalic vein and is designed to transfer tissue from the volar aspect of the forearm. The skin of the flap is incised and dissection is carried down to the underlying fascia. Subfascial dissection is used to elevate the medial and lateral sides of the flap toward the vascular pedicle, leaving a bed of muscle proximally and intact paratenon distally. The pedicle is identified between the flexor carpi radialis and brachioradialis muscles and the radial artery with its vena comitantes are divided distally. The dissection plane is developed under the vascular pedicle along its length and is carried proximally into the antecubital fossa. The recurrent branch of the radial artery is identified and provides the limits of arterial dissection. The flap is then harvested and inset into the defect.

The greatest shortcoming of the RFF is limited surface area, making it unsuitable for resurfacing larger defects. Postoperative tendon exposure is not uncommon, but is almost always self-limiting and resolves with basic wound care. There is also the potential for hand ischemia following sacrifice of the radial artery. Fortunately, this complication is extremely rare and can be avoided with a preoperative Allen's test [26].

Case Report – Radial Forearm Flap

A 56-year-old male presented with a history of squamous cell carcinoma of the right external auditory canal. He underwent resection with canal wall down mastoidectomy and obliteration of the external auditory canal followed by adjuvant radiation therapy. The tumor recurred and was found to involve the squamous temporal bone to the underlying dura, as well as the temporalis muscle and surrounding soft tissue. A lateral temporal bone resection was performed in conjunction with a parotidectomy and right neck dissection. The resulting defect included a large area of exposed dura (fig. 1). Reconstruction was performed with an RFF flap (fig. 2, 3). His postoperative hospital course was uncomplicated.

Latissimus Dorsi Flap

The latissimus dorsi muscle is a broad, flat muscle that originates from the thoracic spinous processes and posterior iliac crest and inserts into the posterior portion of the humerus. It is supplied by the thoracodorsal vessels, which originate from the subscapular system. Virtually the entire muscle may be utilized as a free flap, making it ideal for reconstructing large skull base defects.

The skin paddle is designed in an oblique fashion over the muscle. The initial incision is made along the anterior aspect of the skin paddle and the anterior leading edge of the muscle is identified. The flap is elevated off the thorax from its multiple origins in a cephalad direction toward the axilla. The vascular pedicle is identified and segmental perforating vessels are ligated sequentially. During the dissection toward the humeral insertion, the serratus anterior muscle and its vascular pedicle are identified. Dividing these provides the longest vascular pedicle for the latissmus flap [1, 25].

The primary disadvantage of this flap is the need to harvest with the patient in the lateral decubitus position. This prevents the more efficient and time-saving efforts of a two-team approach. If the serratus anterior is included with the flap, there is a potential for winging of the scapula and chronic shoulder pain. Additionally, the latissimus dorsi is one of the thinnest muscles in the body and the majority of the flap's tissue bulk is the muscle itself. This atrophies following denervation, which may compromise its ability to fill dead space and provide stable, long-term contouring [26].

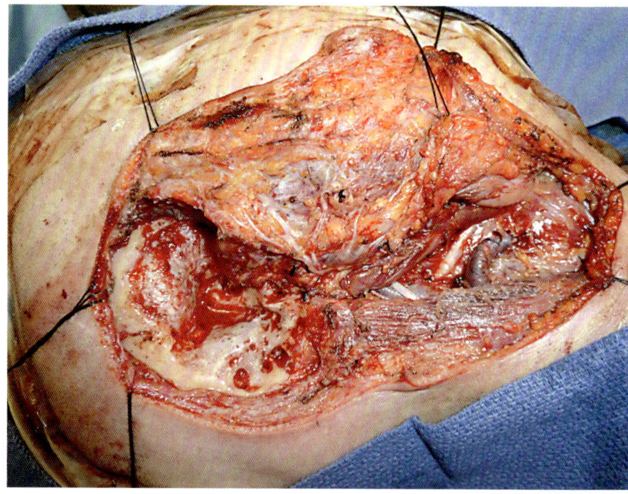

Fig. 1. Ablative defect with exposed dura.

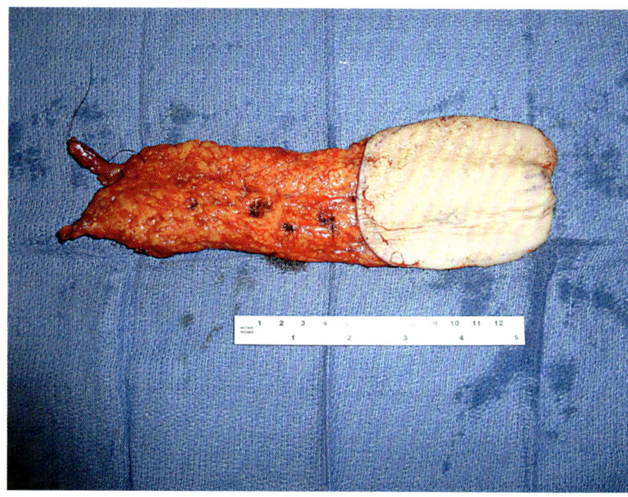

Fig. 2. Radial forearm free flap.

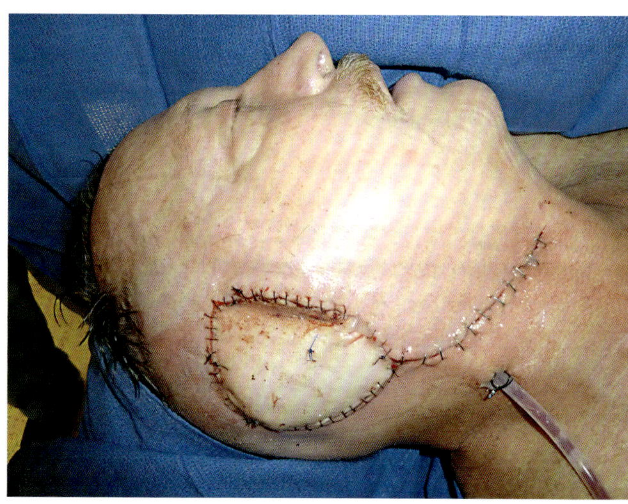

Fig. 3. Reconstruction after inset of the RFF flap.

Anterolateral Thigh Flap

The ALT flap is a versatile flap with a long vascular pedicle and low donor site morbidity. It has enjoyed widespread use in Asia for almost 30 years and subsequently gained increasing popularity in Europe and North America over the past decade [27–30]. In 2006, Malata et al. [23] reported a series of eight patients who underwent ALT flap reconstruction for lateral skull base defects. All flap transfers were successful. The reconstructive and donor site morbidities were minimal. Hanasono et al. [31] echoed these findings in their review of 34 patients who underwent ALT flap reconstruction for anterior, lateral, and posterior skull base defects. The primary disadvantage of the ALT flap is the anatomic variability of the perforating vessels. This has become less problematic as experience with the flap harvest increases and by taking a cuff of vastus lateralis muscle to incorporate the musculocutaneous perforators.

With the patient in a supine position, a line is drawn between the anterior superior iliac spine and the superolateral border of the patella. This line represents the muscular septum between the rectus femoris and vastus lateralis muscles. The cutaneous vessels are mapped by handheld Doppler centered over the midpoint of this line, with the majority of skin perforators within a 3-cm radius of this midpoint. The skin paddle is centered over these vessels, with the long axis parallel to that of the thigh. Dissection begins at the medial border of the flap over the rectus femoris muscle. An incision is made through the deep fascia and the flap is raised laterally until the intermuscular septum is reached. The descending branch of the lateral femoral circumflex artery and, if present, perforating septocutaneous vessels are identified in the groove between the rectus femoris and vastus lateralis muscles. Once the entire course and suitability of the pedicle is confirmed, the lateral skin incision is made. A small cuff of fascia is maintained around the septocutaneous perforators and a 2-cm wide cuff of vastus lateralis muscle is harvested with the flap in order to preserve the musculocutaneous perforators to the skin paddle [32]. The pedicle is then skeletonized, ligated and the flap harvest completed.

Case Report – Anterolateral Thigh Flap

A 6-year-old female presented with recurrent osteosarcoma of the right orbit involving the anterior and lateral skull base. She underwent composite resection with orbital exenteration, total maxillectomy and infratemporal fossa resection to the foramen ovale. Dissection extended to the dura at the skull base (fig. 4, 5). Reconstruction was performed with an ALT flap, which provided dural coverage, obliterated the ablative dead space, and allowed for relining of the nasal cavity and creation of a hemi-palate (fig. 6, 7). She underwent revision of her venous anastomosis on post-operative day one, but had an otherwise uncomplicated hospital course without wound or CNS complications.

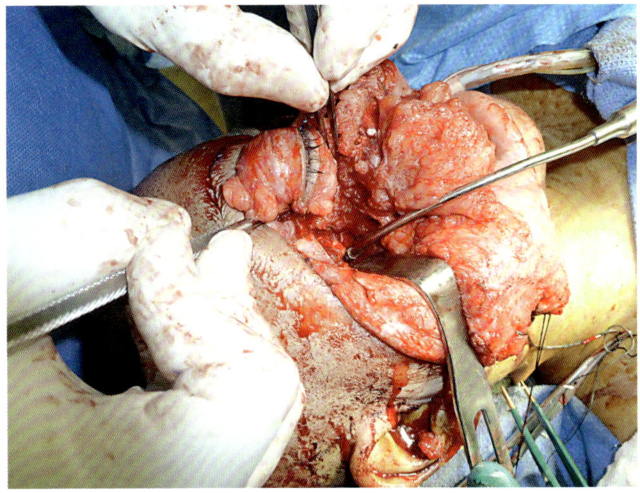

Fig. 4. Intraoperative view of the orbital exenteration with tumor visible in the maxillary sinus.

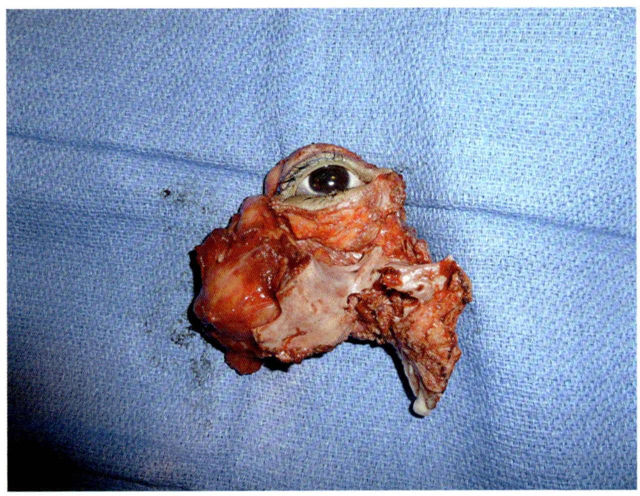

Fig. 5. Composite resection with orbital exenteration and total maxillectomy.

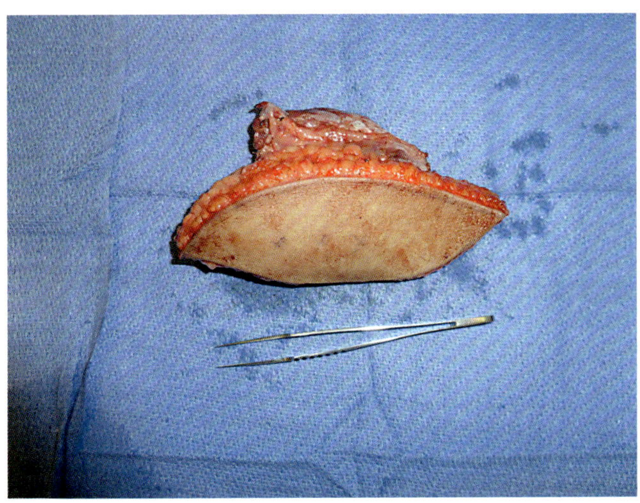

Fig. 6. Harvested ALT.

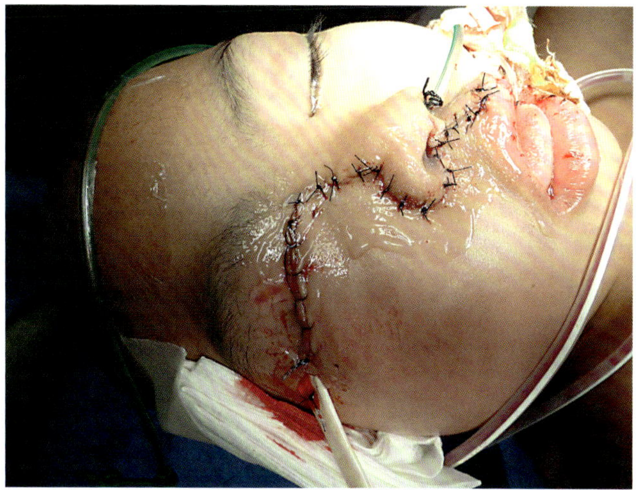

Fig. 7. Closure following microvascular reconstruction.

Postoperative Management and Complications

The incidence of major local and/or systemic postoperative complications following microvascular reconstruction of the skull base ranges from 30 to 40% [7, 9–11, 14, 15]. Early complications (<28 days) include: partial or total flap loss, CSF leak, hematoma, facial nerve weakness, pneumocephalus, myocardial infarction, deep vein thrombosis, pulmonary embolus, blindness and seizure. Late complications (>28 days) consist of palatal fistula, delayed wound healing or wound infection, meningitis, intracranial abscess, infected hardware, ectropion, enophthalmos, orbital dystopia, persistent diplopia, and facial cellulitis [1, 24]. Wound and CNS-related complications occur most frequently [8, 11, 15, 24]. Medical comorbidities and prior treatment with radiation therapy are statistically significant predictors of wound complications. Significant risk factors for CNS complications include prior radiation therapy, dural invasion, and brain invasion [8].

With success rates approximating 95%, free flaps provide a reliable reconstructive option for skull base defects [10, 11, 15, 24]. As noted previously, the use of free tissue transfer following craniofacial resection results in a significant reduction of post-operative complications, especially CNS-related complications [14, 15]. CSF leaks occur less frequently and typically resolve without surgical intervention. The reported incidence of meningitis and intracranial abscess is also decreased [10, 24]. Despite these advances, Ganley et al. [8] still note a postoperative mortality of 4.7% following craniofacial resection. The only independently significant risk factor for death is the presence of medical comorbidities. In the setting of technical innovations and improved peri-operative care, the stability of mortality rates is likely explained by the surgeon's increasing ability to perform more complex tumor resections and address more challenging reconstructive dilemmas.

Conclusion

Over the past two decades, microvascular free tissue reconstruction has reduced complications and improved outcomes in skull base surgery. The complexity of these defects demands that reconstructive surgeons be facile with multiple techniques and flap designs. Four of the most commonly used flaps – the RAM, RFF, latissimus dorsi, and ALT – collectively provide surgeons with the means to address a diverse range of defects. Despite our increasing expertise, postoperative morbidity and mortality rates remain high. Thus, these patients demand careful preoperative selection and optimization, as well as meticulous postoperative care.

References

1 Pusic AL, Chen CM, Patel S, et al: Microvascular reconstruction of the skull base: a clinical approach to surgical defect classification and flap selection. Skull Base 1997;17: 5–15.

2 Kraus DH, Shah JP, Arbit E, Galicich JH, Strong EW: Complications of craniofacial resection for tumors involving the anterior skull base. Head Neck 1994;16:307–312.

3 Deschler DG, Gutin PH, Mamelak AN, et al: Complications of anterior skull base surgery. Skull Base 1996;6:111–117.

4 Dias FL, Sa GM, Klingerman J, et al: Complications of anterior craniofacial resection. Head Neck 1999; 21:12–20.

5 Donald PJ: Complications in skull base surgery for malignancy. Laryngoscope 1999;109:1959–1966.

6 Solero CL, DiMeco F, Sampath P, et al: Combined anterior craniofacial resection for tumors involving the cribiform plate: early postoperative complications and technical considerations. Neurosurgery 2000;47:1296–1304.

7 Patel SG, Singh B, Polluri A, et al: Craniofacial surgery for malignant skull base tumors: report of an international collaborative study. Cancer 2003;98: 1179–1187.

8 Ganley I, Patel SG, Singh B, et al: Complications of craniofacial resection for malignant tumors of the skull base: report of an international collaborative study. Head Neck 2005;27:455–451.

9 Disa JJ, Rodriguez VM, Cordeiro PG: Reconstruction of lateral skull base oncological defects: the role of free tissue transfer. Ann Plast Surg 1998; 41:633–639.

10 Newman J, O'Malley BW, Chalian A, Brown MT: Microvascular reconstruction of cranial base defects: an evaluation of complication and survival rates to justify the use of this repair. Arch Otolaryngol Head Neck Surg 2006;132:381–384.

11 Teknos TN, Smith JC, Day TA, et al: Microvascular free tissue transfer in reconstructing skull base defects: lessons learned. Laryngoscope 2002;112: 1871–1876.

12 Katz MR, Irish JC, Devins GM, et al: Psychosocial adjustment in head and neck cancer: the impact of disfigurement, gender, and social support. Head Neck 2003;25:103–112.

13 Maclean DH, Buncke HJ: Autotranplant of omentum to a large scalp defect with microsurgical revascularization. Plast Reconstr Surg 1972;49:268–274.

14 Neligan PC, Mulholland S, Irish J, et al: Flap selection in cranial base reconstruction. Plast Reconstr Surg 1996;98:1159–1168.

15 Califano J, Cordeiro PG, Disa JJ, et al: Anterior cranial base reconstruction unsing free tissue transfer: changing trends. Head Neck 2003;25:89–96.

16 Georgantopoulou A, Hodgkinson PD, Gerber CJ: Cranial-base surgery: a reconstructive algorithm. Br J Plast Surg 2003;56:10–13.

17 Chang DW, Langstein HN, Gupta A, et al: Reconstructive management of cranial base defects after tumor ablation. Plast Reconstr Surg 2001;107: 1346–1355.

18 Izquierdo R, Leonetti JP, Origitano TC, et al: Refinements in using free-tissue transfer from complex cranial base reconstruction. Plast Reconstr Surg 1993;92:567–574.

19 Clayman GL, DeMonte F, Jaffe DM, et al: Outcome and complications of extended cranial-base resection requiring microvascular free-tissue transfer. Arch Otolaryngol Head Neck Surg 1995;121:1253–1257.

20 Aspoas AR, Wilson GR, McLean NR, et al: Microvascular reconstruction of complex craniofacial defects. Ann R Coll Surg Engl 1997;79:278–283.

21 Djalilian HR, Gapany M, Levine SC: Reconstruction of complicated skull base defects utilizing free tissue transfer. Skull Base 2002;12:209–213.
22 Reza Nouraei SA, Ismail Y, Gerber CJ, et al: Long-term outcome of skull base surgery with microvascular reconstruction for malignant disease. Plast Reconstr Surg 2006;118:1151–1158.
23 Malata CM, Tehrani H, Kumiponjera D, et al: Use of anterolateral thigh and lateral arm fasciocutaneous free flaps in lateral skull base reconstruction. Ann Plast Surg 2006;57:169–175.
24 Chiu ES, Kraus D, Bui DT, et al: Anterior and middle cranial fossa skull base reconstruction using microvascular free tissue techniques: surgical complications and functional outcomes. Ann Plast Surg 2008;60:514–520.
25 Urken ML, Blackwell KE: Rectus abdominis; in Urken ML, Cheney ML, et al (eds): Regional and Free Flaps for Head and Neck Reconstruction. Philadelphia, Lippincott Williams & Wilkins, 2012, pp 141–161.
26 Schmalbach CE, Webb DE, Weitzel EK: Anterior skull base reconstruction: a review of current techniques. Curr Opin Otolaryngol Head Neck Surg 2010;18:238–243.
27 Wei FC, Jain V, Ortho MC, et al: Have we found an ideal soft-tissue flap? An experience with 672 anterolateral thigh flap. Plast Reconstr Surg 2002;109: 2219–2226.
28 Makitie AA, Beasley NJP, Neligan PC, et al: Head and neck reconstruction with anterolateral thigh flap. Otoloaryngol Head Neck Surg 2003;129: 547–555.
29 Yu P: Characteristics of the anterolateral thigh flap in a western population and its application in head and neck reconstruction. Head Neck 2004;26: 759–769.
30 Lin DT, Coppit GL, Burkey BB: Use of the anterolateral thigh flap for reconstruction of the head and neck. Curr Opin Otolaryngol Head Neck Surg 2004; 12:300–304.
31 Hanasono MM, Sacks JM, Goel N, et al: The anterolateral thigh free flap for skull base reconstruction. Otolaryngol Head Neck Surg 2009;140:155–160.
32 Lueg EA: The anterolateral thigh flap: radial forearm's 'big brother' for extensive soft tissue head and neck defects. Arch Otolaryngol Head Neck Surg 2004;130:813–818.

Derrick T. Lin, MD
Skull Base Center, Massachusetts Eye and Ear Infirmary
243 Charles Street
Boston, MA 02114 (USA)
E-Mail derrick_lin@meei.harvard.edu

Bleier BS (ed): Comprehensive Techniques in CSF Leak Repair and Skull Base Reconstruction.
Adv Otorhinolaryngol. Basel, Karger, 2013, vol 74, pp 92–103 (DOI: 10.1159/000342284)

Spontaneous Cerebrospinal Fluid Leak and Management of Intracranial Pressure

Zachary M. Soler · Rodney J. Schlosser

Department of Otolaryngology-Head and Neck Surgery, Medical University of South Carolina, Charleston, S.C., USA

Abstract
Cases of cerebrospinal fluid (CSF) rhinorrhea without a clear etiology are known as spontaneous CSF leaks. Spontaneous CSF leaks most commonly occur in the ethmoid skull base or lateral sphenoid and rarely stop without surgical intervention. This chapter reviews the clinical presentation, diagnostic workup, and treatment of patients with spontaneous CSF rhinorrhea. Specific attention will be given to endoscopic methods of repair and the importance of idiopathic intracranial hypertension as a possible comorbidity and contributing factor.

Cases of cerebrospinal fluid (CSF) rhinorrhea without a clear etiology, such as history of traumatic injury, are known as spontaneous CSF leaks. This chapter will review the clinical presentation and management strategy for spontaneous CSF rhinorrhea, with a focus on endoscopic methods of repair. Specific attention will be given to the importance of idiopathic intracranial hypertension (IIH) as a contributing factor. The pathophysiology of IIH will be reviewed, including diagnostic workup and current treatment paradigms.

Spontaneous CSF Leaks

Most CSF leaks have a clear etiology, such as those secondary to acute traumatic accidents, congenital skull base dehiscence, or expected outcome of a tumor resection. Those leaks without an identifiable inciting event or cause are usually considered 'spontaneous'. In early case series, spontaneous CSF leaks accounted for only 3–5% of all leaks [1–4]. However, more recent case series report much higher rates for spontaneous leaks, ranging from 14 to 46% [5–7]. Although modern surgical repair methods can achieve >90% success for CSF leaks as a whole, the outcomes reported for spontaneous leaks were historically much worse [8]. When examining surgical failures by etiology, spontaneous leaks accounted for a majority of all failures. In comparison to a 10% failure rate for all leaks, initial studies estimated failure rates after repair of spontaneous CSF leaks between 25 and 87% [9–11]. The apparent relative increase in spontaneous CSF leaks, together with their increased proclivity toward recurrence, generated a renewed interest into their pathophysiology and clinical management.

Pathophysiology

Historically, there was little distinction between 'spontaneous' and 'idiopathic' CSF leaks. However, over the last decade significant attention has been given to the role of elevated intracranial pressure (ICP) as a contributing factor for spontaneous leaks [12–14]. Recent series focusing solely on cases of spontaneous leakage have consistently documented increased ICP in those undergoing surgical repair, with average ICP ranging from 24 to 26.5 cm H_20 (normal 0–15 cm H_20) [13, 15]. Most patients with spontaneous CSF leaks actually fit diagnostic criteria for IIH, a condition of unexplained elevated ICP formerly known as psuedotumor cerebri or benign intracranial hypertension [16, 17]. The current hypothesis is that dural pulsations generated from elevated ICP exert direct pressure on the bony skull base. Over time, the thinnest areas of the skull base would completely erode, allowing brain and meninges to herniate and CSF to leak. This hypothesis presumes both elevated ICP and a skull base predisposed to erosion, often secondary to hyperpneumatization.

Clinical Features

Clear rhinorrhea is the defining clinical feature of a sinonasal CSF leak. In an upright position, clear fluid will drain anteriorly or in a postnasal fashion depending on the site of the leak. In most cases, placing the head in a dependent position increases the rate of leakage and often the sidedness is readily apparent. Because spontaneous leaks lack a clear inciting event, symptoms are often initially attributed to more common conditions such as allergic rhinitis and correct diagnosis may be delayed. Headaches are seen in roughly 44–91% of spontaneous CSF leak cases, with patients also commonly reporting secondary symptoms such as imbalance (13.2–27%), tinnitus (13.2–45%), and visual disturbance (10.5–18%) [15–17]. Many of these secondary symptoms persist after definitive repair, suggesting that they are related more closely to the elevated ICP than the leak itself. Lastly, spontaneous CSF leaks can predispose

Table 1. Modified Dandy criteria for IIH

Diagnostic criteria (must satisfy all criteria)
1. Signs and symptoms of increased ICP (headaches, nausea, vomiting, visual changes, papilledema)
2. Elevated CSF pressure on lumbar puncture (≥25 cm H_20 opening pressure)
3. Normal CSF chemical/cytologic composition
4. Absence of localizing or focal neurologic signs (excludes cranial nerve 6 palsy)
5. Exclusion of cerebral venous thrombosis on neuroimaging (usually requires CT scanning or MRI)

ICP = Intracranial pressure; CSF = cerebrospinal fluid; CT = computed tomography; MRI = magnetic resonance imaging.

to meningitis, and in these rare cases patients may present with classical symptoms such as fever, photophobia, and cervical stiffness.

Demographics

Patients presenting with spontaneous CSF leak are typically middle-aged, overweight women [15, 18, 19]. Although published series include a few patients who are young, thin, and/or male, these cases are the exception. Most case series are overwhelmingly comprised of women, ranging from 77 to 85%. The average age across studies is between 50 and 61 years and the vast majority has an elevated BMI. In a study of 56 patients undergoing surgical repair, Woodworth et al. [18] found that 82% had an elevated BMI averaging 36.2. This same level of obesity was seen in separate series reported by Seth et al. [15] and Schlosser et al. [19]. In fact, the clinical characteristics of spontaneous CSF leak patients closely mirror those seen in IIH, with regard to age, gender, and BMI. In the Schlosser et al. [17] series, 72% of patients met the modified Dandy criteria for IIH – leading the authors to speculate that many spontaneous CSF leaks may be an end result of IIH.

Idiopathic Intracranial Hypertension

IIH is a syndrome characterized by elevated ICP of unknown cause. The Modified Dandy Criteria have been proposed to classify those with IIH and include clinical signs/symptoms of increased ICP, elevated CSF pressure on lumbar puncture, and

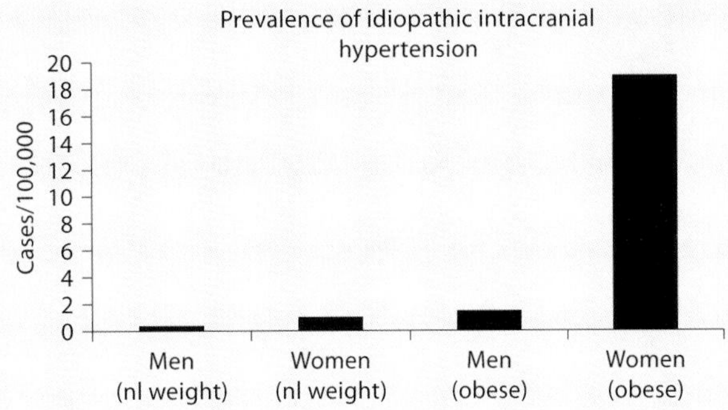

Fig. 1. Epidemiology of IIH.

lack of localizing or focal neurologic signs (table 1) [20]. To fit these criteria, patients must also demonstrate normal CSF chemical composition and demonstrate lack of cerebral venous thrombosis on imaging. IIH occurs most commonly in obese women of child-bearing age. The prevalence of IIH is approximately 1 case/100,000 women, but this increases 13-fold in women aged 20–44 who are overweight and 19-fold in women who are obese (fig. 1) [21, 22]. The prevalence is much less in males (0.3 cases/100,000), but, as in women, obese men are 5 times more likely to have IIH than men of normal weight. Studies have also shown that men with IIH are more likely to have obstructive sleep apnea and twice as likely to develop severe visual loss [23].

At present the pathophysiologic cause of IIH is poorly understood. Theories include reduced CSF absorption, increased CSF production, and increased venous pressure secondary to stenosis of large cerebral venous sinuses [24]. This latter theory is based on the fact that CSF exits through arachnoid granulations into the venous sinuses, such that a downstream narrowing of a large venous sinus could predispose to elevated ICP. Narrowing of venous sinuses is consistently found in patients with IIH; however, it is not clear whether it represents a cause, effect or both via a positive feedback loop. Regardless of theory, each must account for the increased prevalence seen in obese women, which has led researchers to explore the role of sex hormones, cortisol dysregulation, and the endocrinologic role of adipose tissue [25].

Patients with IIH suffer important reductions in quality-of-life (QOL) and are more likely to be depressed than age-matched peers. Reduction in QOL reflects symptoms secondary to elevated ICP, including headache, tinnitus, balance problems, and nausea/vomiting. However, visual changes represent the major morbidity associated with IIH. Papilledema is the most common ocular manifestation of IIH and occurs in nearly all patients. Up to 25% of IIH patients may progress to optic atrophy and permanent visual loss [26]. Oftentimes, peripheral vision is compromised first, followed by more central visual fields. Loss of vision and associated

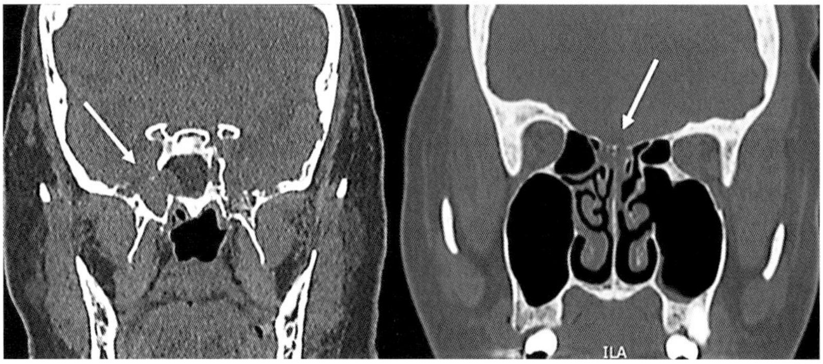

Fig. 2. Left: Coronal CT scan showing encephalocele of lateral right sphenoid sinus. Right: Coronal CT scan showing erosion of ethmoid roof and likely site of CSF leak.

visual symptoms can be subtle and for this reason formal ophthalmologic examination is critical. The economic impact of IIH is estimated at over USD 400 million per year in the United States alone, a figure representing both direct medical costs and lost productivity [27].

Imaging of Spontaneous CSF Leaks

Confirmation of a spontaneous CSF leak is similar to that used to diagnose CSF leaks in general. The diagnostic algorithm has been detailed in earlier chapters of this work and includes appreciation of the clinical picture, sinonasal endoscopy, beta-2-transferrin testing, and imaging.

Assuming a leak is objectively confirmed, the goals of imaging include (1) identifying the site of leak, (2) operative planning, and (3) recognizing signs of elevated ICP/IIH. Computed tomography (CT) imaging is usually the initial procedure of choice, as it provides superior detail of the bony skull base and careful examination will often provide clues as to the possible location of the leak. Magnetic resonance imaging provides important complementary information, including the presence of a meningoencephalocele and/or signs of elevated ICP.

Location and Operative Planning

CT scanning can give important clues as to the location of a spontaneous CSF leak (fig. 2). In many instances, a thinning or dehiscence of the skull base is visible, occasionally with surrounding thickened mucosa or dependent pooling of fluid. The most common location for a spontaneous leak varies by case series, but generally involves the lateral aspect of the sphenoid, cribriform plate, or ethmoid roof. A series of 55 spontaneous leaks reported by Woodworth et al. [18] found

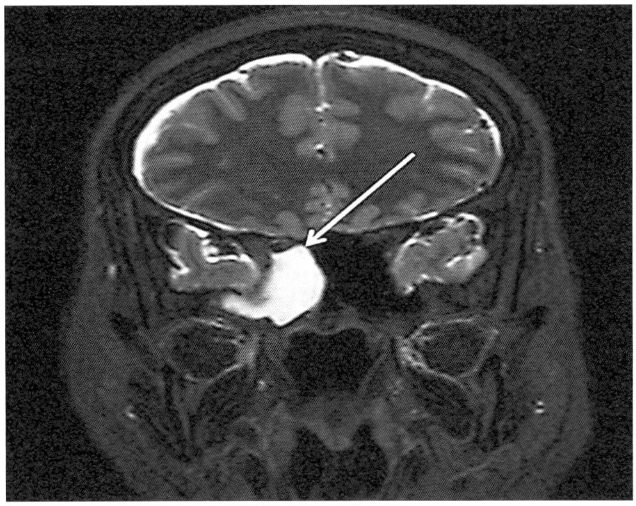

Fig. 3. T_2 MRI scan of meningocele of lateral right sphenoid sinus.

42% in the lateral recess of the sphenoid (usually when hyperpneumatized), 31% in the ethmoid roof, and 22% in the cribriform plate. Seth et al. [15], in a series of 39 patients, found leaks to be most common in the cribriform plate (51%), followed by the lateral sphenoid recess (31%) and ethmoid roof (8%). Lateral sphenoid CSF leaks have been hypothesized to occur at the lateral craniopharyngeal canal (Sternberg's canal), but recent large series using high-resolution CT scan have called this theory into question [28]. Central sphenoid and frontal sinus are relatively rare locations for spontaneous leaks. The possibility of more than one site of leakage must also be considered, with studies reporting multiple sites 3–16% of the time [15, 18].

The majority of spontaneous CSF leaks will have an associated encephalocele or meningoencephalocele (85–96%) [8]. MRI scan can accurately discriminate fluid from soft tissue, allowing the nature and extent of the encephalocele to be estimated (fig. 3). Prior to surgery, the surgeon can usually have a clear picture of the skull base defect which will be encountered and thus can plan the repair and counsel the patient appropriately.

Signs of Increased ICP on Imaging

Patients with spontaneous CSF leaks will commonly have radiologic evidence of elevated ICP. The most common sign is an empty or partially empty sella turcica. A study by Schlosser and Bolger [19] found that 15/15 patients had an empty sella on imaging with 10 completely empty and 5 partially empty. An empty sella turcica most commonly occurs when elevated ICP causes dura to herniate through the sellar diaphragm into the sella turcica, filling it with CSF. The dural herniation and CSF compress the normal pituitary gland against the walls of the sella. Less commonly, empty

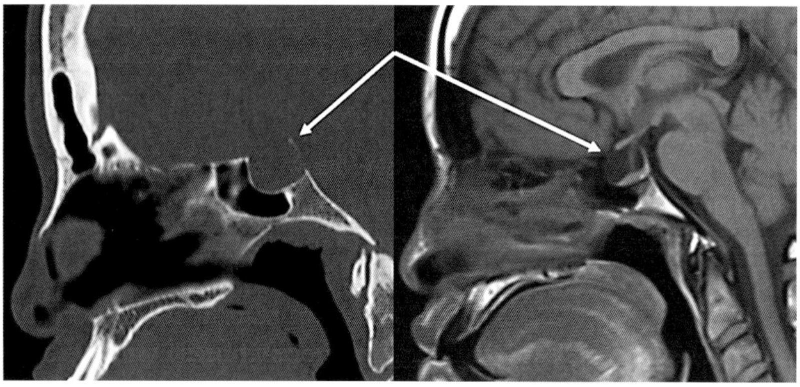

Fig. 4. CT scan (left) and T_1 MRI (right) showing empty sella turcica.

sella may be idiopathic or secondary to pituitary atrophy/necrosis. On MRI scan, T_1-weighted images of the sella will be hypointense compared to brain and T_2 images will be hyperintense compared to brain, following the signaling characteristics of CSF (fig. 4). Patients with IIH may also have additional imaging characteristics secondary to elevated ICP, such as dilated optic nerve sheaths, posterior globe flattening, or stenosis of cerebral venous sinuses [26].

Endoscopic Repair of Spontaneous CSF Leaks

Surgical repair of CSF leaks and skull base defects is covered in other articles in this series and many techniques apply to spontaneous leaks. Nuances specific to spontaneous leaks usually reflect the belief that elevated ICP contributes to the pathophysiology of the leak and must be considered in the operative and perioperative management plan [18]. Failure to consider the implications of increased ICP may account for the high failure rates seen following early attempts at surgical repair. Our surgical management begins with placement of lumbar drain, which allows fluorescein to be used to localize the leak intraoperatively and ensure a watertight seal following repair. The lumbar drain is left open following surgery, reducing pressure at the repair site during the early phase of healing. Use of lumbar drains is based on theoretical premises and controlled clinical trials will likely never be available to evaluate their effectiveness. Because lumbar drains may result in complications, their use should be considered on an individual basis.

The next important factor is choice of repair material and method. Considering that elevated ICP is likely to persist in the postoperative period, we lean towards using underlay grafting techniques and rigid or semirigid materials. Ideally, bone or cartilage can be shaped to span the defect and placed in an underlay fashion. This material is meant to buttress the brain and meninges, minimizing the effect of dural pulsations. The ease by which a rigid underlay can be accomplished varies based on location and patient-specific anatomy, but is usually hardest to achieve in the cribriform plate, where surrounding bone is thin and intracranial space narrow. In very narrow defects (<2 mm), a rigid underlay may not be practical and a wedge shaped graft may be utilized, similar to corking a bottle. Mucosal overlay grafts are usually placed over the rigid graft, and these can be free or pedicled. Mucosal grafts are secured into position using absorbable gelatin paste and/or oxidized cellulose. Nasal sponges are than placed to support the repair and minimize turbulent airflow from sneezing, coughing, and nose-blowing. These are removed between 24 and 48 h after surgery at the surgeon's discretion. The lumbar drain is opened at the time of repair and used to remove 5–10 ml CSF/h for the first 24–48 h.

Placement of a permanent CSF drain, in the form of a ventriculoperitoneal shunt (VPS) or lumboperitoneal shunt (LPS) can be used in very select cases. Permanent shunts are generally considered in patients who have failed multiple, technically adequate repairs or develop additional leaks at different sites. Prior series report use of a permanent shunt in 4–15% of cases; however, these series come from tertiary medical centers and no definitive protocols exist to determine which patients require this additional step [15, 18].

Transpterygoid Approach to Lateral Sphenoid Recess

The lateral recess of the sphenoid sinus is a common location for spontaneous CSF leaks. Leaks in this region generally are associated with a hyperpneumatized sinus, extending lateral to the vidian canal and foramen rotundum [14]. Skull base defects in this region can be large and encephaloceles represent herniation of the temporal lobe into the sinus, often in close proximity to the optic nerve and carotid artery. Given the lateral location of these defects, a standard transnasal sphenoidotomy usually provides insufficient access, even with angled scopes and instruments. The transpterygoid approach typically provides superior visualization and access for repair (online suppl. video 1).

The transpterygoid approach begins with a standard maxillary antrostomy and transethmoid sphenoidotomy. The natural ostium of the sphenoid sinus is enlarged circumferentially, taking care to extend it as far lateral as possible. The maxillary antrostomy is enlarged maximally in a posterior direction until hard palatine bone is encountered. The mucosa is then incised vertically and elevated in a posterior direction until the crista ethmoidalis is identified. This bone is present in >95% of cases and marks the location of the sphenopalatine artery posteriorly [29]. A Kerrison ronguer is then used to remove the posterior wall of the maxillary sinus from medial to lateral beginning with the crista ethmoidalis. The periosteum of the pterygopalatine fossa (PPF) is bipolared and incised horizontally superior to the course of the internal maxillary artery (IMA). Using blunt dissection, the neural and vascular contents of the PPF are retracted until the posterior wall is reached (anterior wall of lateral sphenoid recess). If necessary, branches of the IMA can be clipped and divided. At this point it is helpful to confirm location using image guidance if available. A curette or high speed drill is then used to enter through the anterior wall of the sphenoid sinus, with the opening expanded circumferentially using Kerrison ronguers. The skull base defect and encephalocele should be directly visible using a zero-degree endoscope and standard instrumentation. Repair can then proceed as described above.

Assessment of Intracranial Pressure

Patients with elevated ICP should be managed medically following surgery. The diagnosis of elevated ICP can be made directly via lumbar puncture or indirectly based on the presence of papilledema or radiologic evidence of increased ICP [30]. The diagnosis of ICP via lumbar puncture is usually based on an elevated opening pressure. However, the opening pressure will be unreliable if CSF is already leaking through the skull base. This has led some surgeons to monitor continuous ICP following repair, assuming that pressures should reach a steady state after normal CSF volumes are replenished [31]. Others have advocated for a separate lumbar puncture in the postoperative period in those patients where the diagnosis is unclear. In our practice, a patient is assumed to have elevated ICP if there is radiologic evidence, most commonly an empty sella. Obese patients with normal imaging are suspected of having elevated ICP/IIH and these patients are typically referred to an ophthalmologist or neurologist for further evaluation.

Medical Management of IIH

Patients suspected of having elevated ICP and IIH should have a formal eye examination and be followed by an ophthalmologist. As mentioned previously, visual deficits can be subtle at first and peripheral vision may be lost permanently before it becomes noticeable to the patient. Treatment of visual loss secondary to IIH is beyond the scope of this article, but can involve medical treatment, optic nerve fenestration, shunting procedures (VPS or LPS), and more recently cerebral venous sinus stents.

The most important medical treatment of IIH is weight loss. Studies consistently report improvement in ICP following low calorie diets. Sinclair et al. [32] report outcomes from a prospective cohort study of 25 women with IIH placed on a low calorie (425 kcal/day) for 3 months. Findings 3 months after completion of the diet showed an average weight loss of 16 kg and a mean decrease in ICP of 8.0 cm H_20. There was also significant improvement in headache scores, optic disk elevation (papilledema), diplopia, and tinnitus.

Evaluation and treatment of obstructive sleep apnea (OSA) is also an important component of IIH treatment. Because most IIH patients are obese, OSA is a frequent comorbidity, particularly in men [23]. Some data even suggests that ICP may transiently increase during apneic events, further contributing to the burden of dural pulsations that may predispose to spontaneous CSF leaks [31]. Treatment of OSA, obesity, and IIH thus are complementary.

Acetazolomide is the most common medication used to treat IIH. As a carbonic anhydrase inhibitor, acetazolomide reduces CSF production thereby decreasing CSF pressures. Despite being considered standard treatment, little high quality data exists supporting its use. A single small study compared azetazolomide to topiramate using an alternate allocation strategy [33]. Both treatment groups showed improved visual fields after a year, but no difference was seen between groups. Most recently, a pilot study randomized patients with IIH to acetazolamide or placebo and followed symptoms, body weight, visual function, and QOL [34]. Unfortunately, the study had problems with recruitment and compliance and finished severely underpowered to show a clinical difference.

In patients with spontaneous CSF leaks suspected of ICP and IHH, we recommend weight loss via diet modification and exercise. Diet modification can begin immediately following diagnosis, with light exercise begun one month following surgery and heavy exercise at 3 months. All patients are questioned for signs and symptoms of OSA and referred for polysomnography as appropriate. Those patients with objective evidence of elevated ICP (either on imaging, lumbar puncture, or fundoscopic exam) are begun on acetazolamide 500 mg b.i.d. Electrolytes are checked periodically, with drops in carbon dioxide anticipated. Medication is discontinued if and when ICP can be controlled via weight loss. Long-term compliance with medication can be problematic, as paresthesias, altered taste, and lethargy may be attributed to the medication [26].

Outcomes after Endoscopic Repair of Spontaneous CSF Leaks

Most outcomes data from surgical repair of CSF leaks comes from small and medium-sized retrospective case series. We recently performed a systematic review with meta-analysis of studies reporting success rates after endoscopic repair. A total of 49 studies were examined detailing 1,326 leak repairs. The summary success rate for all types of CSF leaks was 90.6% (95% CI 80–92.5%). Interestingly, the success rate for spontaneous CSF leak repairs appears to have increased over time. When examining data from 7 series published within the last 5 years, the summary success rate was 91.1% (95% CI 86.7–94.6%). The increase in success over time may reflect an improved understanding of the importance of elevated ICP and adoption of comprehensive management strategies, such as those outlined above.

References

1 Lanza DC, O'Brien DA, Kennedy DW: Endoscopic repair of cerebrospinal fluid fistulae and encephaloceles. Laryngoscope 1996;106:1119–1125.

2 Mattox DE, Kennedy DW: Endoscopic management of cerebrospinal fluid leaks and cephaloceles. Larygoscope 1990;100:857–862.

3 Zlab MK, Moore GF, Daly DT, Yonkers AJ: Cerebrospinal fluid rhinorrhea: a review of the literature. Ear Nose Throat J 1992;71:314–317.

4 Burns JA, Dodson EE, Gross CW: Transnasal endoscopic repair of cranionasal fistulae: a refined technique with long-term follow-up. Laryngoscope 1996;106:1080–1083.

5 Kirtane MV, Gauitham K, Upadhyaya SR: Endoscopic CSF rhinorrhea closure: our experience in 267 cases. Otolaryngol Head Neck Surg 2005;132: 208–212.

6 Carrau RL, Snyderman CH, Kassam AB: The management of cerebrospinal fluid leaks in patients at risk for high-pressure hydrocephalus. Laryngoscope 2005;115:205–212.

7 Mirza S, Tahaper A, McClelland L, Jones NS: Sinonasal cerebrospinal fluid leaks: management of 97 patients over 10 years. Laryngoscope 2005;115: 1774–1777.

8 Wise SK, Schlosser RJ: Evaluation of spontaneous nasal cerebrospinal fluid leaks. Curr Opin Otolaryngol Head Neck Surg 2007;15:28–34.

9 Hubbard JL, McDonald TJ, Pearson BW, Laws ER: Spontaneous cerebrospinal fluid rhinorrhea: evolving concepts in diagnosis and surgical management based on the Mayo Clinic experience from 1970 through 1981. Neurosurgery 1985;16:314–321.

10 Schick B, Ibing R, Brors D, Draf W: Long-term study of endonasal duraplasty and review of the literature. Ann Otol Rhinol Laryngol 2001;110: 142–147.

11 Gassner HG, Ponikau JU, Sherris DA, et al: CSF rhinorrhea: 95 consecutive surgical cases with long term follow-up at the Mayo Clinic. Am J Rhinol 1999;13:439–447.

12 Schlosser RJ, Bolger WE: Significance of empty sella in cerebrospinal fluid leaks. Otolaryngol Head Neck Surg 2003;128:32–38.

13 Schlosser RJ, Wilensky EM, Grady S, Bolger WE: Elevated intracranial pressures in spontaneous cerebrospinal fluid leaks. Am J Rhinol 2003;17:191–195.

14 Woodworth BA, Palmer JN: Spontaneous cerebrospinal fluid leaks. Curr Opin Otolaryngol Head Neck Surg 2009;17:59–65.

15 Seth R, Rajasekaran K, Luong A, Benninger MS, Batra PS: Spontaneous CSF leaks: factors predictive of additional interventions. Laryngoscope 2010;120: 2141–2146.

16 Yang Z, Wang B, Wang C, Liu P: Primary spontaneous cerebrospinal fluid rhinorrhea: a symptom of idiopathic hypertension? J Neurosurg 2011;115: 165–170.

17 Schlosser RJ, Woodworth BA, Wilensky EM, Grady MS, Bolger WE: Spontaneous cerebrospinal fluid leaks: a variant of benign intracranial hypertension. Ann Otol Rhinol Laryngol 2006;115:495–500.

18 Woodworth BA, Prince A, Chiu AG, et al: Spontaneous CSF leaks: a paradigm for definitive repair and management of intracranial hypertension. Otolaryngol Head Neck Surg 2008;138: 715–720.

19 Schlosser RJ, Bolger WE: Spontaneous nasal cerebrospinal fluid leak and empty sella syndrome: a clinical association. Am J Rhinol 2003;17:91–96.

20 Fraser C, Plant GT: The syndrome of psuedotumor cerebri and idiopathic intracranial hypertension. Curr Opin Neurol 2011;24:12–17.

21 Radhakrishnan K,Ahlskog JE, Cross SA, Kurland LT, O'Fallon WM: Idiopathic intracranial hypertension (psuedotumour cerebri): descriptive epidemiology in Rochester, Minn. Arch Neurol 1993;50: 78–80.

22 Durcan FJ, Corbett JJ, Wall M: The incidence of psuedotumour cerebri population studies in Iowa and Louisiana. Arch Neurol 1988;45:875–877.

23 Bruce BB, Kedar S, Van Stavern GP, et al: Idiopathic intracranial hypertension in men. Neurology 2009; 72:304–309.

24 De Simone R, Ranieri A, Bonavita V: Advancement in idiopathic intracranial hypertension pathogenesis: focus on sinus venous stenosis. Neurol Sci 2010;31(suppl 1):S33–S39.

25 Sinclair AJ, Ball AK, Burdon MA, Clarke CE, Stewart PM, Curnow SJ, Rauz S: Exploring the pathogenesis of IHH: an inflammatory perspective. J Neuroimmunol 2008;201–202:212–220.

26 Bruce BB, Biousse V, Newman NJ: Update on idiopathic intracranial hypertension. Am J Ophthalmol 2011;152:163–169.

27 Friesner D, Rosenman R, Lobb BM, Tanne E: Idiopathic intracranial hypertension in the USA: the role of obesity in establishing prevalence and healthcare costs. Obes Rev 2011;12:e372–e380.

28 Baranano CF, Cure J, Palmer JN, Woodworth BA: Sternberg's canal: fact or fiction? Am J Rhinol Allergy 2009;23:167–171.

29 Bolger WE, Borgie RC, Melder P: The role of the crista ethmoidalis in endoscopic sphenopalatine artery ligation. Am J Rhinol 1999;13:81–86.

30 Schlosser RJ, Wilensky EM, Grady MS, Palmer JN, Kennedy DW, Bolger WE: Cerebrospinal fluid pressure monitoring after repair of cerebrospinal fluid leaks. Otolaryngol Head Neck Surg 2004;130: 443–448.
31 Reh DD, Gallia G, Ramanathan M, et al: Perioperative continuous cerebrospinal fluid pressure monitoring in patients with spontaneous cerebrospinal fluid leaks: presentation of a novel technique. Am J Rhinol 2010;24:238–243.
32 Sinclair AJ, Burdon MA, Nightingale PG, et al: Low energy diet and intracranial pressure in women with idiopathic intracranial hypertension: prospective cohort study. BMJ 2010;340:c2701.
33 Celebisoy N, Gokcay F, Sirin H, Ayurekli O: Treatment of idiopathic intracranial hypertension: topiramate vs. acetazolamide, an open-label study. Acta Neurol Scand 2007;116:322–327.
34 Ball AK, Howman A, Wheatley K, et al: A randomized controlled trial of treatment for idiopathic intracranial hypertension. J Neurol 2011;258:874–881.

Rodney J. Schlosser, MD
Department of Otolaryngology-Head and Neck Surgery, Medical University of South Carolina
171 Ashley Avenue
Charleston, SC 29425 (USA)
E-Mail schlossr@musc.edu

OSM

Bleier BS (ed): Comprehensive Techniques in CSF Leak Repair and Skull Base Reconstruction.
Adv Otorhinolaryngol. Basel, Karger, 2013, vol 74, pp 104–118 (DOI: 10.1159/000342285)

Endoscopic Reconstruction of Frontal, Cribiform and Ethmoid Skull Base Defects

David Chin · Richard J. Harvey

University of New South Wales, Macquarie University, St Vincent's Hospital, and Macquarie University Hospital, Sydney, Australia; Changi General Hospital, Singapore, Singapore

Abstract

Endoscopic skull base reconstruction (ESBR) is an established option for complete reconstruction of the anterior skull base (ASB). While free mucosal grafts are still relevant for limited defects (<10 mm), the evolution of vascular flaps has extended the limits of reconstruction with up to 94% success in defect closure and avoidance of cerebrospinal fluid (CSF) leak. This article discusses the planning and execution in ESBR for anterior fossa defects. Factors influencing the choice of reconstruction (flap location, size, influence of raised intracranial pressure, high flow CSF leak, radiotherapy) are examined. The reconstructive options are discussed, including the nasoseptal flap (NSF), inferior turbinate pedicled flap and the endopericranial flap, with particular emphasis on the technical aspects of reconstruction using the NSF. A step-by-step description, with an accompanying video, is provided. Techniques applicable to ESBR in general (subdural graft placement, graft/flap placement, barrier and support dressing) are covered. Novel flaps for ASB reconstruction are also briefly-described.

Techniques for endoscopic skull base reconstruction (ESBR) have their origins in repairs for small, traumatic or spontaneous cerebrospinal fluid (CSF) leaks. The aim of reconstruction is to provide a robust separation of the cranial and sinonasal cavities to prevent intracranial infection [1–3]. Endoscopic repair of anterior skull base (ASB) defects avoids the morbidity of open approaches, in particular, of brain retraction, anosmia and craniotomy. The fundamental basis for traditional transcranial approaches was for the introduction of a vascular pericranial flap to assist with repair. A variety of endoscopic vascular repair techniques are now well described [1, 4]. In a systematic review of ESBR, the success rate for vascular flaps in preventing CSF leaks was 94% [5]. ESBR is applicable even after failed open reconstruction [6]. Conversely, the success rate with free mucosal grafts is lower [5, 7, 8]. Thus, they are only suitable for small defects without unfavorable prognostic factors [1]. ESBR has thus evolved from open approaches using nonautologous material and regional myofascial flaps [9] to endoscopic approaches using vascular local flaps [5, 8, 10, 11]. Robust ESBR has facilitated the ability to manage an increasingly expansive variety of skull base pathologies over the last decade. The challenge in ESBR is the increasing size of defects [12] and an operative field often exposed to previous radiation therapy for malignancy [13–15].

Evaluation of the Skull Base Defect

The critical factors in the evaluation of a defect are its location on the skull base, its size and the degree of arachnoid disruption. Additionally, the presence of comorbidities known to influence success of the repair, in particular, raised intracranial pressure (ICP) [16, 17], prior radiotherapy [1, 15] and persistent pneumocephalus [18], need to be assessed.

Location
For pedicled vascular flaps, the location of the skull base defect in relation to the origin of the supplying vessel is an important factor. The nasoseptal flap (NSF) reaches most ASB defects [19, 20] and is the workhorse of ESBR. Alternatives flaps in the fronto-ethmoidal region require a favorable arc of rotation and these are discussed in the following sections.

Size
Small defects (<10 mm) generally do not need repair with vascular flaps [21]. Guidelines for repair based on the dimensions of a defect evolved to describe the limits of nonvascularized options. These concepts are still valid in the management of small skull base defects. With the NSF, a defect of almost any length in the ASB can be repaired when the flap is harvested to include the nasal floor and the anterior septal wall right up to the nasal vestibule. Laterally, its width is sufficient for a complete ethmoid roof defect extending from one orbital wall to the other [19] (fig. 1).

Arachnoid Disruption and Cisternal Exposure
In the ASB, defects which extend into the arachnoid cisterns or third ventricle can result in a high flow CSF leak, which is associated with higher failure rate (6%) compared to low flow states (2%) [7, 11]. CSF diversion should be considered in these situations.

Raised Intracranial Pressure
Elevated intracranial pressure (ICP) is a key factor in spontaneous CSF leaks [16]. While benign intracranial hypertension and empty sella syndrome are uncommon, postoperative high-pressure hydrocephalus secondary to subarachnoid hemorrhage

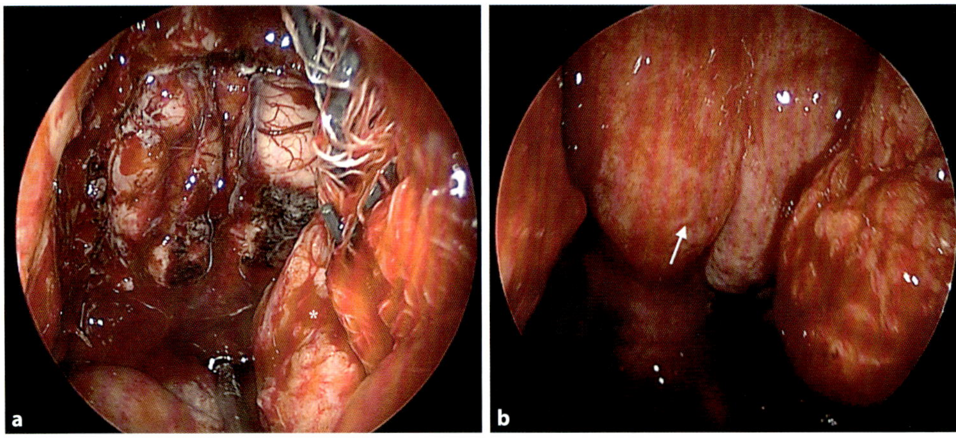

Fig. 1. **a** Complete anterior skull base defect combined with resection of left medial orbital wall (asterisk) for squamous cell carcinoma of the left ethmoid cavity. **b** NSF fully covering the defect (white arrow).

and cerebral edema can lead to flap disruption. If elevated ICP exists [17], measures ranging from temporary [22] to permanent [23] CSF diversion and the use of acetazolamide to decrease CSF production [24] should be considered.

Special Considerations for Fronto-Ethmoidal Defects

The ideal positioning of subdural graft material requires an adequate circumferential bony shelf for support. However, for benign unilateral pathologies, there may be an attempt to preserve olfaction on the normal side by performing unilateral ethmoid roof resection without removal of the crista galli. In this situation, there is no horizontal bony ledge medially and the medial flap edge runs superiorly against the falx cerebri.

In malignant disease, bilateral olfactory fossa resection with removal of the crista galli (fig. 2) allows the subdural graft to be supported on the contralateral ethmoid roof. The argument against bilateral resection in clinically unilateral malignancy assumes that no contralateral tumor spread has occurred and that the some olfaction can be preserved. However, the concept that the thin superior septum can provide a reliable barrier to tumor spread is tenuous at best. In reality, the likelihood of preserving contralateral olfaction after adjuvant radiotherapy is small compared with the benefit of a wide sinonasal cavity after complete ethmoid roof resection.

Preparation before Reconstruction

A wide approach is required to facilitate surgical access, flap positioning and postoperative sinonasal function. Complete sphenoethmoidectomy and marsupialization of the sphenoid sinus creates a smooth bed over which the flap can be draped.

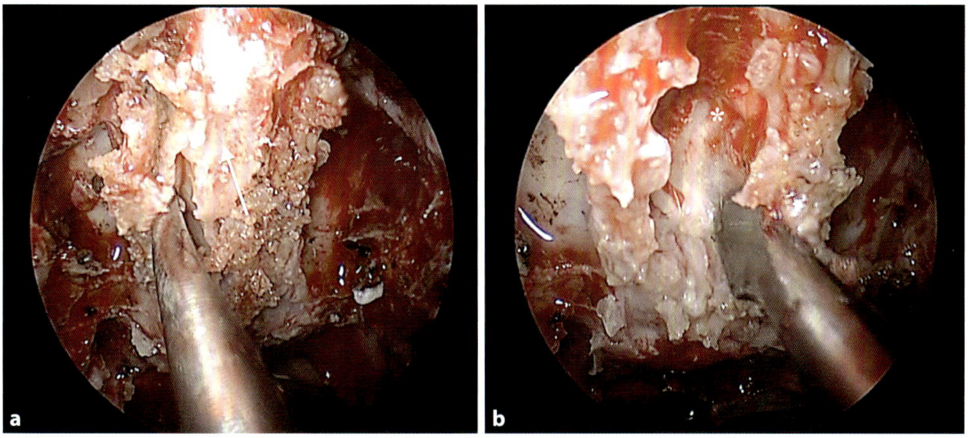

Fig. 2. a Intra-operative photo of crista galli (white arrow) being dissected. **b** Crista galli has been drilled away from between the falx cerebri (asterisk).

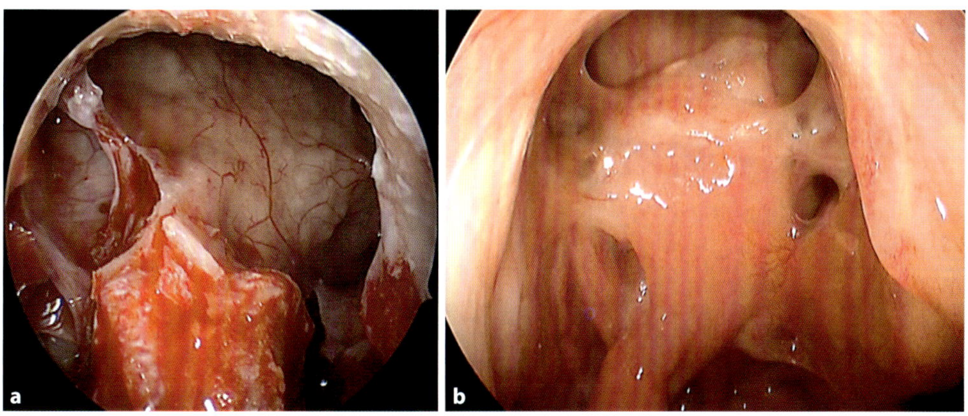

Fig. 3. a Intraoperative picture of Draf 3 approach in a case of endoscopic tumor resection. **b** Postoperative picture of Lothrop cavity with functional mucosa and good access for tumor surveillance.

A Draf 3 frontal sinusotomy is used when the reconstruction involves areas anterior to the anterior ethmoidal artery. We have a low threshold for performing a Draf 3 frontal sinusotomy [25] as this facilitates flap placement and ensures a functioning frontal cavity with direct access for tumor surveillance (fig. 3).

An area surrounding the defect is demucosalized to avoid the separation of the flap or graft from the recipient bed due to trapped mucus secretions. Deliberate removal of mucosa beneath the pedicle of the flap is required but does not need to be comprehensive as there is no evidence of increased mucocele formation from trapped mucosal remnants under the flap [26]. Osteoneogenesis may be stimulated by light abrasion with a diamond burr or curette.

Techniques describing intracranial, extradural placement of bone or cartilage layers necessitate circumferential dural elevation around the defect. However, the belief that rigid support is required to prevent cerebral herniation is not supported by the literature; none of a case series of 120 patients developed cerebral herniation after skull base reconstruction without rigid support despite the fact that these defects subtended a large surface area [27].

Intracranial Layer

Traditional reports were of trying to place an intracranial extradural layer. This is neither practical nor necessary and thus the intracranial layer is placed in a subdural fashion.

Subdural Reconstruction

Various materials for dural replacement are available. The subdural graft should be 10 mm larger than the defect circumferentially. The authors' preference is to use a dural graft matrix (Duragen®, Integra Life Sciences Corporation, N.J., USA). Alternative materials include autologous fat [13, 22], fascia lata [3, 27] or allografts such as acellular dermis [28], lyophilized dura [3, 29], Ethisorb™ (Codman, Mass., USA) [30], LactoSorb (Biomet Microfixation, Fla., USA) and Tutoplast (Tutogen Medical GmbH, Germany) [31]. Occasionally, more than one layer can be used.

The olive-tipped antral suction probe (without suction attached) or straight curette is ideal for manipulating the subdural graft; ball tip probes too often pinch the graft material and are not ideal. The graft should have no folds when properly-positioned, allowing the overlapping portions of the graft to be compressed between the dura and the brain. No CSF should be visibly leaking around the graft (fig. 4). Suturing of subdural grafts is not required although it has been described [32].

Extracranial Layer

Free Mucosal Graft

Free mucosal grafts have been used successfully for small (<10 mm) skull base defects associated with CSF leak repair and endoscopic skull base surgery (ESBS) [21, 22, 33]. The inferior turbinate can provide a graft of reasonable size with minor donor site morbidity. The turbinate is divided along its attachment to the lateral nasal wall

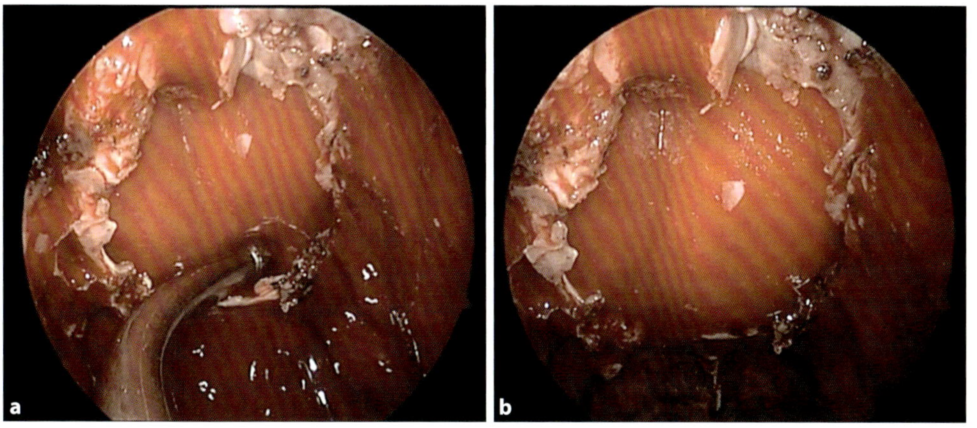

Fig. 4. **a** Using the olive-tipped suction to maneuver the subdural graft. **b** Correctly positioned subdural graft with no folds and no CSF leak.

as in a turbinectomy procedure. Preparation of the graft is done ex vivo with 'butterflying' of the mucosal flaps and removal of the turbinate bone (fig. 5). Releasing incisions are made on the periosteal layer to allow the graft to conform to the contour of the recipient site. If other donor sites are considered, posterior septal mucosa should not be used as this compromises the availability of the NSF for future use.

Vascular Flaps

Nasoseptal Flap

The NSF has a robust vascular supply via the posterior septal branch of the sphenopalatine artery. Studies comparing dimensions of the donor site in relation to anticipated skull base resections suggest that the NSF can cover any defect confined to the ASB [19, 20], but may be inadequate for more extensive defects, e.g. spanning from sella to frontal sinus [19].

Technique
Mucosal incisions are made using monopolar diathermy with a bent needle-point tip. An initial back-cut is made on the vertical plate of palatine bone and along the posterior choana to improve pedicle mobilization for coverage of anterior defects. This continues anterior to the posterior edge of the nasal septum, to the nasal floor and then laterally into the inferior meatus. The incision can be extended up to the junction with the skin of the vestibule to maximize anterior coverage (fig. 6). Incorporating the mucosa of the hard palate greatly increases the width of the flap and is preferable to harvesting from the superior septum.

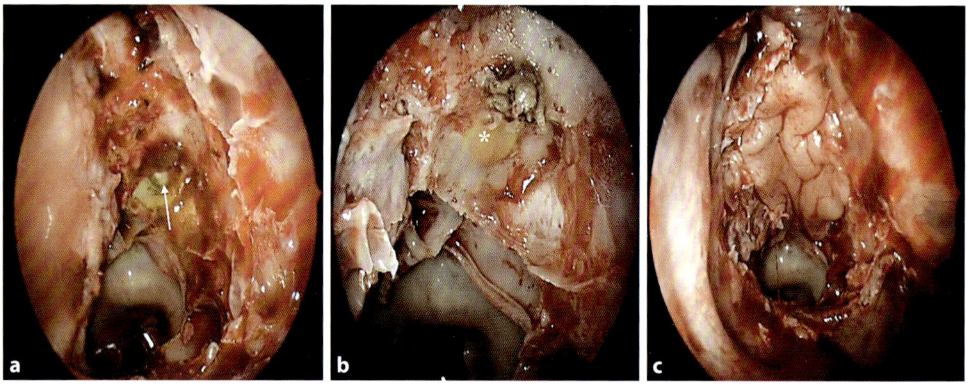

Fig. 5. Repair of traumatic CSF leak of the ethmoid skull base. **a** Intraoperative picture of defect exposed with fluorescein-stained CSF (white arrow). **b** Subdural graft in place (asterisk). **c** Free mucosal graft (inferior turbinate) has been 'filleted' and laid over defect.

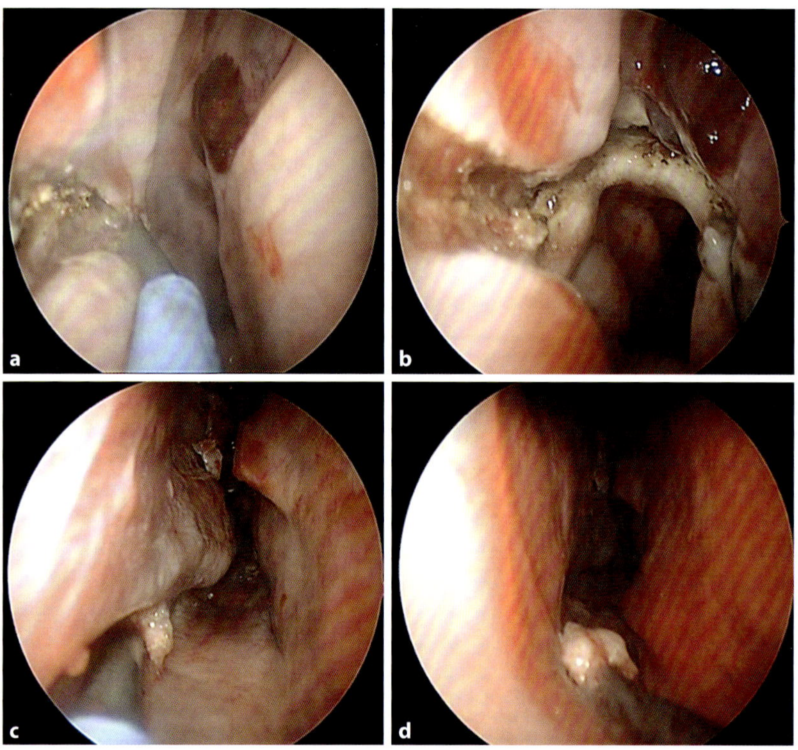

Fig. 6. Lower incision for the NSF. **a** Initial back cut on the posterior fontanelle. **b** Incision continuing on to the posterior choana and then anterior to the free edge of septum. **c** Cut on the nasal floor out to the floor of the inferior meatus. **d** Cut continued anteriorly to the vestibule.

A second incision begins at the medial aspect of the sphenoidotomy, which has been previously performed with care to avoid injury to the pedicle inferiorly. It continues forward at the level of the inferior edge of the superior turbinate (ST) and middle turbinate (MT). Anterior to the MT, the incision turns superiorly to include the swell body area (another technique to increase width and gain better exposure) and is connected with the first incision anteriorly (fig. 7a–d).

There are important reasons for maintaining the height of the second incision at the level of the ST and MT. For malignancies, the superior strip of mucosa contralateral to the tumor cannot be included in the flap and the absence of tumor in the patient margin should be verified on frozen section histopathology. For benign lesions, sparing of mucosa above the level of the ST and MT preserves olfactory mucosa and minimizes postoperative hyposmia. Furthermore, the thin mucoperiosteal layer on the superior septum is not ideal for the flap as it is neither thick nor robust (online suppl. video 1). OSM

Results
The NSF is associated with a success rate of over 90% [5, 7, 8, 14, 34–36]. The success rates are preserved even with NSF takedown and reuse in staged or revision ESBR [37]. Rare cases of flap necrosis have been reported and were usually associated with prior radiation therapy [14, 15].

Limitations
The NSF provides adequate coverage for adults but in pediatric patients under the age of 14, the degree of coverage and success from reconstruction is variable [38]. Septal growth is most rapid between the ages of 10 and 13 and the average potential flap length is less than the average length of the ASB until 9 years of age. Careful preoperative measurements are required if NSF reconstruction is contemplated in this transitional age group.

Modifications
While the use of bilateral NSFs to repair large ASB defects has been reported in a small case series [39], in the opinion of the author, the concept of bilateral flaps does not reflect formal skull base training. No advantage is gained by submaximal septal flap harvesting bilaterally as compared to a single large, long, unilateral flap, taken to the vestibular skin anteriorly and to the inferior meatus laterally including the palatal floor. Reports of 'bilateral' flaps are merely preservation of the mucosa of posterior septal window (at least on one side). Such short flaps are of little use in skull base surgery. Furthermore, bilateral mucosal elevation leaves denuded septal cartilage and bone on both sides of the septum which prolongs the return of nasal mucosal function unless a formal posterior septectomy is also performed.

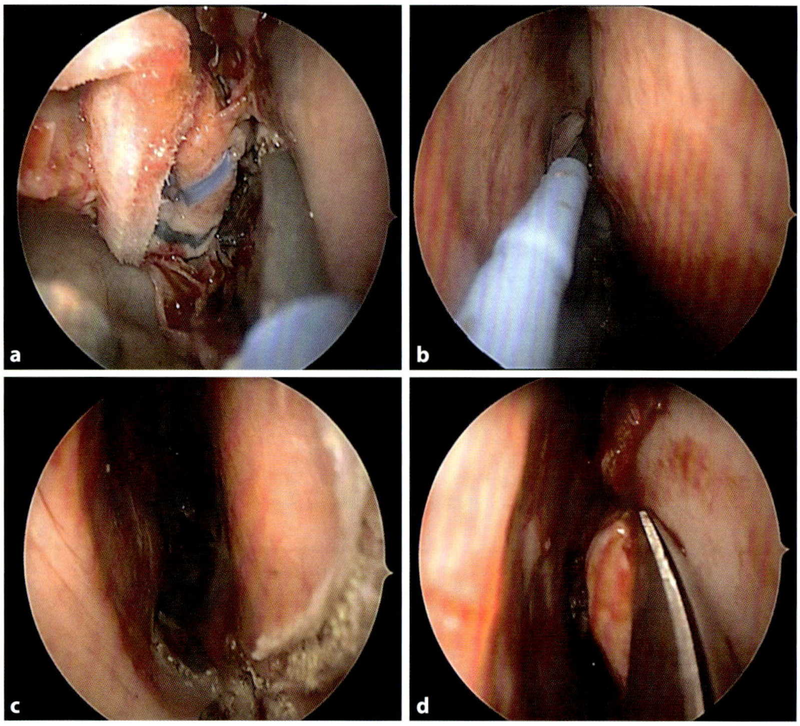

Fig. 7. Upper incision for the NSF. **a** Staying below 'olfactory' mucosa of the nasal septum at the height of inferior border of the superior and middle turbinates. **b** Cut rising superiorly over the septal swell body. **c** Cut descending over the anterior septum to join the lower incision at the vestibule. **d** Superior attachment of previously elevated NSF divided with iridectomy scissors.

Pericranial Flap

The flap has the potential to cover large areas, with a reach that extends to the posterior cranial fossa. The vascular supply is derived from the supraorbital and supratrochlear vessels which are branches of the ophthalmic artery from the internal carotid system. Use of the pericranial flap (PCF) may be considered when a vascularized mucosal flap is unavailable or inadequate for repair of the anticipated defect or in the setting of an endoscopic assisted open approach.

Technique

The PCF is harvested by standard bicoronal incision. The avoidance of a craniotomy and brain retraction are the advantages of this approach even in ESBR. Subgaleal dissection exposes the pericranial layer. The vessels are identified at the orbital rim and a 3-cm pedicle is protected. If a very posterior defect is anticipated, the pericranial incision can be as far posterior as the length of flap required. The PCF is then elevated in the subperiosteal plane. A nasionotomy (5 × 15 mm) is drilled to allow transposition of the flap from its extracranial position into the nasal cavity. The PCF is overlaid on the skull base defect with the pedicle in the midline.

Although described via a brow lift approach with endoscopic flap elevation [35], a 10-mm glabellar incision is still required for creation of the nasionotomy and this approach is rarely performed even in the institution of these reports.

Results
The reliability of the PCF is well established. Success with the minimally invasive endoscopic approach was reported for a case series of 10 patients, 8 of whom underwent postoperative radiation therapy with no adverse effects on the reconstruction [40].

Inferior Turbinate Pedicled Flap

The inferior turbinate pedicled flap (ITPF) is based on the inferior turbinate artery which is derived from the posterior lateral nasal artery, a branch of the sphenopalatine artery [41]. Based on a cadaveric study, the mean reach of the ITPF was 67% of the distance to the anterior limit of the ASB [42]. Consequently, careful preoperative assessment must be used to predict if the ITPF will cover the anticipated defect, especially if the frontal table is involved. Due to the curvilinear shape of the donor mucosa, the harvested flap should be unrolled before being mobilized to cover the skull base defect (fig. 8)

Results
In a report of 4 patients who had ITPF for ESBR, no flap failures were encountered [41]. These patients had had previous posterior septectomy as part of ESBS, hence precluding the use of the NSF.

Novel Flaps

Novel flaps have been reported in an attempt to overcome the limitations of posteriorly based pedicled flaps especially for defects involving the frontal sinus. The success of these innovations remain to be seen as the limits of ESBS are extended and with an increasing number of patients in whom the NSF is not available.

Bipedicled Anterior Septal Flap
An anteriorly based bipedicled septal flap has been described in a radioanatomic and cadaveric study [43]. The vascular supply is from the septal branch of the superior labial artery and the nasopalatine artery. The flap is estimated

to cover all of the frontal beak and about 50% of the posterior frontal table. If combined with a posteriorly based flap, full frontal skull base coverage is achievable.

Anteriorly Based Lateral Nasal Wall Flap

An anteriorly based lateral nasal wall flap has also been described in a report where it was successfully employed to reconstruct defects of the ASB in 3 patients [44].

Anteriorly Based Interior Turbinate Flap

An anteriorly based inferior turbinate flap, based on supply from the anterior ethmoidal artery, was recently described [45]. The angle of rotation allows reconstruction at the posterior frontal table in addition to the cribiform plate. Success of the anteriorly based inferior turbinate flap, alone or in combination with other flaps, was reported in a small series of 7 patients with high-flow intraoperative CSF leak.

Pedicled Facial Buccinator Flap

A mycomucosal flap based on a branch of the facial artery was recently described in cadaveric studies [46, 47]. Transposition of the flap was via maxillotomy without facial incisions and the flap was able to reach the anterior skull base and planum sphenoidale. While further studies are needed, this flap shows promise as a reconstruction alternative.

Flap Placement

Correct flap placement involves complete coverage of the defect with good apposition of the flap layer with the subdural graft and to the defect margin (fig. 9). Sinus walls are marsupialized and bony ledges burred down to allow the flap to contour to the walls of the cavity from the origin of the pedicle to the reconstruction site. After ensuring that there is no leak of CSF, fibrin glue is applied to the edges of the mucosal flap to facilitate adherence of the flap to the skull base. Glue, clot and foreign material between layers are avoided.

Dressings and Support

The reconstruction site is lined with a layer of resorbable dressing, e.g. Gelfoam® (Pfizer, N.Y., USA). Pledgets of compressed Gelfoam (1–2 cm) placed over the flap, serve as a 'barrier' to minimize disruption to the reconstruction when supportive dressings are subsequently inserted, adjusted and finally removed (fig. 10a). Options for support dressings include gloved Merocel® (Medtronic, Conn., USA) sinus packs, Rapid Rhino® (ArthroCare, Tex., USA) inflatable packs or the balloon of a Foley's catheter. If a Foley's catheter is used, the balloon is inflated under endoscopic

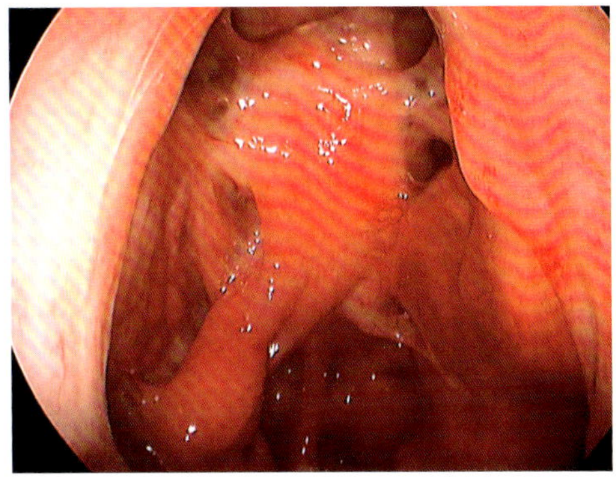

Fig. 8. Post-operative endoscopic view of the ITPF.

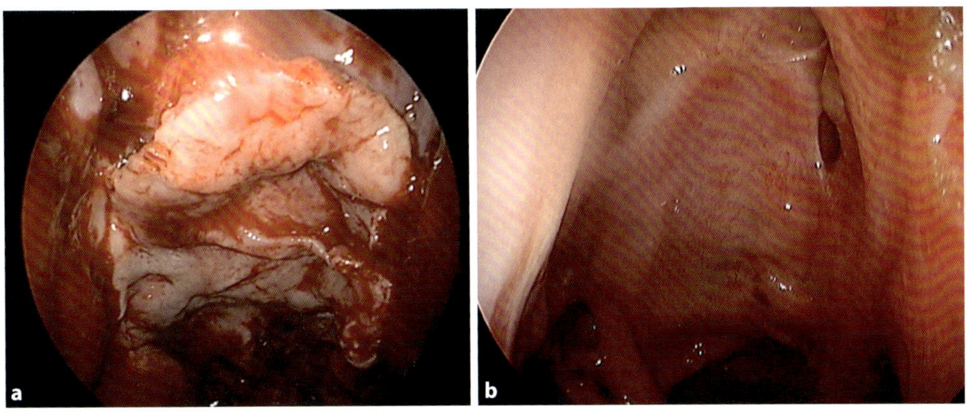

Fig. 9. Intraoperative (**a**) and postoperative (**b**) photo of NSF well positioned over the ethmoid roof defect.

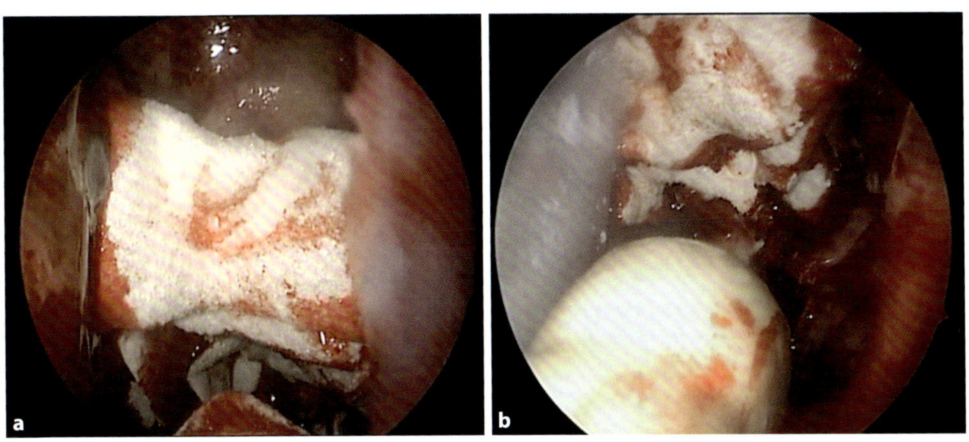

Fig. 10. Dressings. **a** Loose Gelfoam packing overlying flap. **b** Foley's catheter balloon being inflated.

visualization (usually about 20 ml required) (fig. 10b). Overinflation is avoided as this can cause displacement of the reconstruction and/or compression of the vascular pedicle. The balloon may be deflated after a few days and subsequently removed at about 1 week postsurgery.

Conclusion

ESBR of the ASB has evolved from the repair of CSF leaks in the ethmoid roof. This approach can now be employed for complete reconstruction of the entire floor of the anterior cranial fossa. While free mucosal grafts are still relevant for limited defects, the evolution of vascular flaps has extended the limits of reconstruction, especially those associated with tumor resection. The NSF is the workhorse of ASB repair and has proven to be reliable and robust. When it is not available or inadequate, alternative flaps may be used to substitute or complement the NSF. Preoperative planning and perioperative care are important for the success of ESBR.

References

1 Harvey RJ, Smith JE, Wise SK, Patel SJ, Frankel BM, Schlosser RJ: Intracranial complications before and after endoscopic skull base reconstruction. Am J Rhinol 2008;22:516–521.

2 Bernal-Sprekelsen M, Bleda-Vazquez C, Carrau RL: Ascending meningitis secondary to traumatic cerebrospinal fluid leaks. Am J Rhinol 2000;14:257–259.

3 Bernal-Sprekelsen M, Alobid I, Mullol J, Trobat F, Tomas-Barberan M: Closure of cerebrospinal fluid leaks prevents ascending bacterial meningitis. Rhinology 2005;43:277–281.

4 Hegazy HM, Carrau RL, Snyderman CH, Kassam A, Zweig J: Transnasal endoscopic repair of cerebrospinal fluid rhinorrhea: a meta-analysis. Laryngoscope 2000;110:1166–1172.

5 Harvey RJ, Parmar P, Sacks R, Zanation AM: Endoscopic skull base reconstruction of large dural defects: a systematic review of published evidence. Laryngoscope 2012;122:452–459.

6 Clark DW, Citardi MJ, Fakhri S: Endoscopic management of skull base defects associated with persistent pneumocephalus following previous open repair: a preliminary report. Otolaryngol Head Neck Surg 2010;142:820–826.

7 Patel MR, Stadler ME, Snyderman CH, Carrau RL, Kassam AB, Germanwala AV, et al: How to choose? Endoscopic skull base reconstructive options and limitations. Skull Base 2010;20:397–404.

8 Kassam AB, Prevedello DM, Carrau RL, Snyderman CH, Thomas A, Gardner P, et al: Endoscopic endonasal skull base surgery: analysis of complications in the authors' initial 800 patients. J Neurosurg 2011;114:1544–1568.

9 Schmalbach CE, Webb DE, Weitzel EK: Anterior skull base reconstruction: a review of current techniques. Curr Opin Otolaryngol Head Neck Surg 2010;18:238–243.

10 Gagliardi F, Boari N, Mortini P: Reconstruction techniques in skull base surgery. J Craniofac Surg 2011;22:1015–1020.

11 Zanation AM, Thorp BD, Parmar P, Harvey RJ: Reconstructive options for endoscopic skull base surgery. Otolaryngol Clin North Am 2011;44:1201–1222.

12 Harvey RJ, Gallagher RM, Sacks R: Extended endoscopic techniques for sinonasal resections. Otolaryngol Clin North Am 2010;43:613–638.

13 Snyderman CH, Kassam AB, Carrau R, Mintz A: Endoscopic reconstruction of cranial base defects following endonasal skull base surgery. Skull Base 2007;17:73–78.

14 El-Sayed IH, Roediger FC, Goldberg AN, Parsa AT, McDermott MW: Endoscopic reconstruction of skull base defects with the nasal septal flap. Skull Base 2008;18:385–394.

15 Zanation AM, Carrau RL, Snyderman CH, Germanwala AV, Gardner PA, Prevedello DM, et al: Nasoseptal flap reconstruction of high flow intraoperative cerebral spinal fluid leaks during endoscopic skull base surgery. Am J Rhinol Allergy 2009;23: 518–521.
16 Schlosser RJ, Wilensky EM, Grady MS, Palmer JN, Kennedy DW, Bolger WE: Cerebrospinal fluid pressure monitoring after repair of cerebrospinal fluid leaks. Otolaryngol Head Neck Surg 2004;130: 443–448.
17 Schlosser RJ, Bolger WE: Nasal cerebrospinal fluid leaks: critical review and surgical considerations. Laryngoscope 2004;114:255–265.
18 Wise SK, Harvey RJ, Patel SJ, Frankel BM, Schlosser RJ: Endoscopic repair of skull base defects presenting with pneumocephalus. J Otolaryngol Head Neck Surg 2009;38:509–516.
19 Pinheiro-Neto CD, Prevedello DM, Carrau RL, Snyderman CH, Mintz A, Gardner P, et al: Improving the design of the pedicled nasoseptal flap for skull base reconstruction: a radioanatomic study. Laryngoscope 2007;117:1560–1569.
20 Pinheiro-Neto CD, Ramos HF, Peris-Celda M, Fernandez-Miranda JC, Gardner PA, Snyderman CH, et al: Study of the nasoseptal flap for endoscopic anterior cranial base reconstruction. Laryngoscope 2011;121:2514–2520.
21 Harvey RJ, Nogueira JF, Schlosser RJ, Patel SJ, Vellutini E, Stamm AC: Closure of large skull base defects after endoscopic transnasal craniotomy. Clinical article. J Neurosurg 2009;111:371–379.
22 Leong JL, Citardi MJ, Batra PS: Reconstruction of skull base defects after minimally invasive endoscopic resection of anterior skull base neoplasms. Am J Rhinol 2006;20:476–482.
23 Carrau RL, Snyderman CH, Kassam AB: The management of cerebrospinal fluid leaks in patients at risk for high-pressure hydrocephalus. Laryngoscope 2005;115:205–212.
24 Seth R, Rajasekaran K 3rd, Luong A, Benninger MS, Batra PS: Spontaneous CSF leaks: factors predictive of additional interventions. Laryngoscope 2010;120: 2141–2146.
25 Chin D, Snidvongs K, Kalish L, Sacks R, Harvey RJ: The outside-in approach to the modified endoscopic Lothrop procedure. Laryngoscope 2012;122: 1661–1669.
26 Bleier BS, Wang EW, Vandergrift WA 3rd, Schlosser RJ: Mucocele rate after endoscopic skull base reconstruction using vascularized pedicled flaps. Am J Rhinol Allergy 2011;25:186–187.
27 Gil Z, Abergel A, Leider-Trejo L, Khafif A, Margalit N, Amir A, et al: A comprehensive algorithm for anterior skull base reconstruction after oncological resections. Skull Base 2007;17:25–37.
28 Germani RM, Vivero R, Herzallah IR, Casiano RR: Endoscopic reconstruction of large anterior skull base defects using acellular dermal allograft. Am J Rhinol 2007;21:615–618.
29 Golusinski W, Wasniewska E, Kulczynski B: Endoscopic reconstruction of the anterior skull base in cerebrospinal rhinorrhea. Otolaryngol Pol 2003;57:75–79.
30 Seidl RO, Todt I, Ernst A: Reconstruction of traumatic skull base defects with alloplastic, resorbable fleece. HNO 2000;48:753–757.
31 Cavallo LM, Messina A, Esposito F, de Divitiis O, Dal Fabbro M, de Divitiis E, et al: Skull base reconstruction in the extended endoscopic transsphenoidal approach for suprasellar lesions. J Neurosurg 2007;107:713–720.
32 Gardner P, Kassam A, Snyderman C, Mintz A, Carrau R, Moossy JJ: Endoscopic endonasal suturing of dural reconstruction grafts: a novel application of the U-Clip technology. Technical note. J Neurosurg 2008;108:395–400.
33 El-Banhawy OA, Halaka AN, Altuwaijri MA, Ayad H, El-Sharnoby MM: Long-term outcome of endonasal endoscopic skull base reconstruction with nasal turbinate graft. Skull Base 2008;18:297–308.
34 Hadad G, Bassagasteguy L, Carrau RL, Mataza JC, Kassam A, Snyderman CH, et al: A novel reconstructive technique after endoscopic expanded endonasal approaches: vascular pedicle nasoseptal flap. Laryngoscope 2006;116:1882–1886.
35 Zanation AM, Snyderman CH, Carrau RL, Kassam AB, Gardner PA, Prevedello DM: Minimally invasive endoscopic pericranial flap: a new method for endonasal skull base reconstruction. Laryngoscope 2009;119:13–18.
36 Horiguchi K, Murai H, Hasegawa Y, Hanazawa T, Yamakami I, Saeki N: Endoscopic endonasal skull base reconstruction using a nasal septal flap: surgical results and comparison with previous reconstructions. Neurosurg Rev 2010;33:235–241.
37 Zanation AM, Carrau RL, Snyderman CH, McKinney KA, Wheless SA, Bhatki AM, et al: Nasoseptal flap takedown and reuse in revision endoscopic skull base reconstruction. Laryngoscope 2011;121:42–46.
38 Shah RN, Surowitz JB, Patel MR, Huang BY, Snyderman CH, Carrau RL, et al: Endoscopic pedicled nasoseptal flap reconstruction for pediatric skull base defects. Laryngoscope 2009;119:1067–1075.

39 Nyquist GG, Anand VK, Singh A, Schwartz TH: Janus flap: bilateral nasoseptal flaps for anterior skull base reconstruction. Otolaryngol Head Neck Surg 2010;142:327–331.
40 Patel MR, Shah RN, Snyderman CH, Carrau RL, Germanwala AV, Kassam AB, et al: Pericranial flap for endoscopic anterior skull-base reconstruction: clinical outcomes and radioanatomic analysis of preoperative planning. Neurosurgery 2010;66:506–12.
41 Fortes FS, Carrau RL, Snyderman CH, Prevedello D, Vescan A, Mintz A, et al: The posterior pedicle inferior turbinate flap: a new vascularized flap for skull base reconstruction. Laryngoscope 2007;117: 1329–1332.
42 Harvey RJ, Sheahan PO, Schlosser RJ: Inferior turbinate pedicle flap for endoscopic skull base defect repair. Am J Rhinol Allergy 2009;23:522–526.
43 Bleier BS, Curry WT, Wang EW, Schlosser RJ: The bipedicled anterior septal flap: a radioanatomic and cadaveric study. Laryngoscopy 2011;121:1367–1371.
44 Hadad G, Rivera-Serrano CM, Bassagaisteguy LH, Carrau RL, Fernandez-Miranda J, Prevedello DM, et al: Anterior pedicle lateral nasal wall flap: a novel technique for the reconstruction of anterior skull base defects. Laryngoscope 2011;121:1606–1610.
45 Gil Z, Margalit N: Anteriorly based inferior turbinate flap for endoscopic skull base reconstruction. Otolaryngol Head Neck Surg 2012;146:842–847.
46 Rivera-Serrano CM, Oliver C, Prevedello D, Gardner P, Snyderman C, Kassam A, et al: Pedicled facial buccinator (FAB) flap: a new flap for reconstruction of skull base defects. Laryngoscope. 2010;120(suppl 4):S234.
47 Rivera-Serrano CM, Oliver CL, Sok J, Prevedello DM, Gardner P, Snyderman CH, et al: Pedicled facial buccinator (FAB) flap: a new flap for reconstruction of skull base defects. Laryngoscope 2010;120:1922–1930.

David Chin
354 Victoria Street
Darlinghurst, NSW 2010 (Australia)
Tel. +61 293604811, E-Mail drdavidchin@gmail.com

Bleier BS (ed): Comprehensive Techniques in CSF Leak Repair and Skull Base Reconstruction.
Adv Otorhinolaryngol. Basel, Karger, 2013, vol 74, pp 119–129 (DOI: 10.1159/000342288)

Planum, Tubercular, Sellar and Clival Defects

Calvin C. Wei · James N. Palmer

Department of Otorhinolaryngology, Head and Neck Surgery, University of Pennsylvania School of Medicine, Hospital of the University of Pennsylvania, Philadelphia, Pa., USA

Abstract

Transsphenoidal and extended transsphenoidal approaches have provided improved surgical access to numerous intradural and extradural tumors from the retrosellar area to the craniovertebral junction. The evolution in reconstruction of planum, tubercular and clival skull base defects has progressed from the use of free tissue grafts alone to the use of free tissue grafts in combination with vascularized flaps. Sellar floor reconstruction is usually necessary only if intraoperative CSF leaks, prolapse of the suprasellar cistern, or bleeding from the medial aspect of the cavernous sinus occur. In patients with intraoperative CSF leak, accurate packing of the sella in the area of the arachnoid membrane tear, with or without packing of the sphenoid sinus, is performed. An increasing armamentarium of local and regional tissue flaps have reduced postoperative CSF leak rates to less than 5%. Future refinements in skull base reconstruction techniques will allow better management of patients at high risk for a postoperative CSF leak, especially those who have been previously irradiated and/or require revision surgery.

Extended endoscopic transsphenoidal approaches have provided improved surgical access to numerous intradural and extradural tumors from the retrosellar area to the craniovertebral junction. Before the adoption of vascularized tissue flaps as a primary reconstructive technique in skull base reconstruction, reconstructive techniques were limited to the use of free tissue grafts. These techniques were adapted from experience accumulated with the endoscopic repair of CSF leaks associated with endoscopic sinus surgery and trauma, and then expanded to repair larger dural defects as well as defects over high-flow intraoperative CSF leaks [1]. As the field of endoscopic skull base surgery has matured, overall CSF leak rates have decreased as a result of improvements in skull base reconstruction utilizing multilayer closure with vascularized local tissue flaps. Zanation et al. [1] analyzed pooled data from a literature search of all articles describing any endoscopic endonasal reconstruction of the skull base from the EMBASE (1980 to December week 1 2010) and Medline (1950 to December week 1 2010) databases. CSF leak rates were stratified into one group using free grafts alone versus the other using vascularized local tissue flaps for reconstruction. The overall rate of CSF leak was 11.5% (70/609). This was represented as 15.6% (51/326) for free grafts and 6.7% (19/283) for vascularized reconstruction. Reconstructive strategies must be modified based on the size and location of the skull base defect, the extent of dural resection, whether the arachnoid is violated, the nature of the patient's CSF pressure, states of poor wound healing such as Cushing's disease and prior irradiation, and whether the patient has had previous surgery or treatment that would preclude use of vascularized local tissue flaps.

Materials for Skull Base Repair

Autologous

Bone

Fragments of autologous bone can be obtained from the middle turbinate, nasal septum and anterior sphenoid wall. Although autologous materials are advantageous for their absent cost and perfect biocompatibility, the amount of useful bone is often unpredictable. Bone fragments have variable and irregular sizes, shapes and thicknesses, and often require the use of more than one fragment which makes the repair cumbersome and unwieldy.

Dura

Abdominal fat or fascia lata have been used for dural repair. Kitano and Taneda [2] have used a double-layered configuration composed of one fragment of fascia and another of expanded polytetrafluoroethylene, which is directly sutured to the dural edges using a modified sphenoid retractor. However, dural borders sufficient for suturing are not always present and tying knots at the defect edges is difficult.

Heterologous

Bone Substitutes

LactoSorb, a copolymer of L-lactic acid and glycolic acid (Walter Lorenz, Jacksonville, Fla., USA), is a bone substitute consisting of a combined copolymer of polylactic acid and polyglycolic acid. LactoSorb is easy to model and completely resorbed within 1 year. It becomes malleable at 70°C, making it easy to shape according to the defect, and becomes rigid after a few seconds at body temperature. It decays to approximately 70% after 8 weeks. The consistency of the implant progressively decreases and is completely resorbed by 1 year [3].

Other materials such as titanium, silicone, and alumina ceramic have also been used as bone substitutes. Titanium is a very inert material that is not associated with a tissue reaction. Silicone is easy to shape, but is often associated with induction of a local inflammatory reaction. Other authors have used hydroxyapatite cement which is directly molded into the osseous defect. However, failure occurs when the material is exposed to fluid during the first 4 h after insertion, the time required for it to become water insoluble. The use of hydroxyapatite in a field that is filled with CSF is not advisable [4]. In addition, there is a high incidence of infection and sequestrum formation associated with these materials [5].

Dural Substitutes

Several dural substitutes are available which include dehydrated human pericardium (Tutoplast, Tutogen Medical GmbH, Neunkirchen am Brand, Germany), collagen matrix (Duragen, Integra Life Sciences, Plainsboro, N.J., USA) and bovine processed collagen (Duraform, Codman, Raynham, Mass., USA). As a dural substitute, human pericardium is thin, not resorbable, and easy to layer. It has been found to effectively combine with LactoSorb. Duraform is a collagen sponge manufactured from processed bovine tendons. It is a sterile, absorbable implant intended for the repair of the dura as a sutureless onlay graft.

Surgical Glues

Several surgical glues are FDA-approved and are described in the literature for use during endoscopic skull base surgery. These include Tisseel/Tissucol (Baxter BioSciences, Deerfield, Ill., USA), BioGlue (CryoLife, Kennesaw, Ga., USA), and DuraSeal (Confluent Surgical, Waltham, Mass., USA). BioGlue is a semisynthetic glue composed of purified bovine serum albumin and glutaraldehyde. While supraphysiologic burst strengths have been reported with BioGlue, glutaraldehyde is associated with some degree of tissue toxicity and application adjacent to critical neurovascular structures should be performed with caution. Tisseel is a two-component biologic system made from pooled human plasma, consisting of fibrinogen solution and thrombin, with a calcium cofactor promoting the final stage of coagulation. These components are deep frozen in preloaded syringes and are mixed during application. The result is a white, elastic mass that adheres firmly to tissue [6]. DuraSeal is a polyethylene glycol hydrogel sealant used for watertight closure of the dura. While DuraSeal is FDA-approved for use in dural closure, it may expand significantly following application and placement in a subdural plane may result in compression of adjacent structures or displacement of an overlay graft.

Sellar Defects

Sellar floor reconstruction is usually necessary only if intraoperative CSF leaks, prolapse of the suprasellar cistern, or bleeding from the medial aspect of the cavernous sinus occur [7, 8]. In the absence of these events, residual space in the sellar cavity may be left open, without any risk of postoperative complications [9]. CSF leaks are caused by localized tears in the arachnoid membrane in the intradural, extracranial space of the sella turcica. Kennedy et al. [10] reported a comprehensive review of a series of transsphenoidal hypophysectomies. With intraoperative CSF leaks, they repaired the sellar defect with fascia or fat held in place by a bone graft. They reported a postoperative leakage rate of 9.6% (11 of 114). Similar outcomes have been reported in other series [11, 12]. Because the dura is easily separated from the bony borders of the sellar floor, an extradural closure is effectively watertight [7]. Reconstruction with a variety of autologous materials including abdominal fat, muscle, cartilage or fascia lata can be performed with simple or stratified packing (intradural and/or extradural). A variety of rigid materials, including cartilage or bone fragments, alumina ceramic, silicone, stainless steel, or titanium plates, have been used to reconstruct the sellar floor [13]. Although there are advantages of autologous materials including biocompatibility and absent cost, the disadvantages, including prolongation of operative time, the need for a separate incision in a predominantly young and female patient population, donor site pain, infection, and hematoma or seroma formation have driven the transition to heterologous materials for sellar repair [8].

Several published series have supported the use of heterologous materials with similar and, in some cases, improved rates of postoperative CSF leak. Lorenz et al. [14] describes the technique of layered reconstruction with acellular dermal allograft, septal bone or cartilage, and acellular dermal allograft placed intracranially with fibrin glue to hold a mucosal free graft over the area of reconstruction. Of 11 patients with intraoperative CSF leaks, 1 patient required secondary closure of a CSF leak. Seiler et al. [13] report on a series of 376 consecutive patients who underwent closure of the sella turcica with a resorbable vicryl patch composed of polyglactin 910/poly-*p*-dioxanone, gelatin foam, and fibrin glue after transsphenoidal surgery. The patch is resorbed at 2–3 months and is replaced by fibrous collagen tissue. They report a postoperative CSF leak rate of 0.5%. Cappabianca et al. [8] describe the use of collagen and/or fibrin sealant to repair intraoperative CSF leaks after endoscopic transsphenoidal pituitary surgery. A piece of collagen fleece is cut to size and placed over the suprasellar cistern and/or the residual pituitary gland. Fibrin sealant is then employed to cover the sellar cavity and/or the sphenoid sinus. When intraoperative CSF leak occurs, collagen fleece and/or fibrin sealant is reinforced by an intra- and/or extradural closure of the sellar floor performed with a polyester-silicone dural substitute (Patch Biomesh N3, Cousin Biotech, Wervicq-Sud, France). They report that of 16 patients who had intraoperative CSF leak repaired with fibrin sealant alone, 6 had a postoperative CSF leak. Of 13 patients who had intraoperative CSF leak repaired with collagen fleece and fibrin sealant, 0 patients had a postoperative CSF leak.

Cappabianca et al. [9] in a series of 170 cases of sellar repair describe their algorithm for sellar and/or sphenoid sinus packing after endoscopic transsphenoidal pituitary surgery. In patients with a microadenoma, macroadenoma without suprasellar extension, a macroadenoma with suprasellar extension that is incompletely removed, no closure or packing of the sella is performed. This is intended to allow descent of any residual tumor in suprasellar tumors. In patients with prolapse of the suprasellar cistern toward the sellar floor without intraoperative CSF leak (e.g. macroadenoma with suprasellar extension, noncommunicating intrasuprasellar arachnoid cyst, intrasuprasellar craniopharyngioma) or bleeding from the medial wall of the cavernous sinus and intracavernous carotid injury, simple intrasellar packing, with intradural and/or extradural closure of the sellar floor, is performed. In patients with intraoperative CSF leak or paninvasive macroadenoma, accurate packing of the sella in the area of the arachnoid membrane tear, with or without packing of the sphenoid sinus, is performed. This procedure consists of a first layer of autologous or synthetic dura, placed intradurally to protect the arachnoid. Autologous fat or a resorbable dural substitute is fitted into the sella. A second layer of autologous or synthetic dura is fitted in the extradural space. The closure is sealed by fibrin glue on the outer surface of the sellar floor. Packing of the sphenoid sinus is useful in the case of a 'ghost' sella or when sellar packing does not look watertight. The mucosa of the sphenoid sinus at the site of repair must be carefully removed. Simple intra- or extradural closure of the sellar floor is sufficient in the case of a suspected but not plainly evident intraoperative CSF leak.

Another paradigm for assessing sellar repair is based on the size and location of the diaphragmatic and/or dural defect. Kelly et al. [15] have categorized intraoperative CSF leaks based on size: no observed leak is categorized grade 0; a small 'weeping' CSF leak, confirmed by the Valsalva maneuver, without a visible diaphragmatic defect is categorized grade 1; a moderate CSF leak with a definite diaphragmatic defect is categorized grade 2, and a large diaphragmatic and/or dural defect is categorized grade 3. Schwartz and coworkers [16] describe their algorithm for reconstruction after endoscopic pituitary surgery based on these categories. For grade 0 defects, nothing is placed in the sella; only a collagen sponge is placed to reduce sellar dead space. For grade 1 defects, a collagen sponge is placed over the exposed suprasellar cistern. The sellar cavity is then filled with fibrin glue (Tisseel, Baxter BioSciences, Deerfield, Ill., USA) and intra- or extradural closure of the sellar floor is performed with a bone substitute (LactoSorb, Lorenz Surgical, Jacksonville, Fla., USA). For grade 2 defects, after a small amount of fibrin glue is injected through the arachnoid defect, layers of collagen sponge are placed over the cisternal surface. The sellar floor is then closed extradurally with a bone substitute. Grade 3 defects more often occur in the setting of extended approaches for suprasellar lesions and require multilayer closure with rigid reconstruction overlying a dural substitute placed in the extradural space. In our experience, rigid reconstructive materials (i.e. bone substitutes) are not routinely used with grade 1 or grade 2 defects, except in cases of high-flow CSF leakage. A lumbar drain is placed if the patient has intraoperative CSF leakage and the closure was not judged to be absolutely watertight or if minimal unexpected postoperative CSF leakage occurs [9].

Planum and Tubercular Defects

The extended transsphenoidal approach expands operative exposure beyond the sella by removing the tuberculum sellae and the planum sphenoidale [17]. This approach is utilized for suprasellar lesions, including pituitary adenomas, tuberculum sella meningiomas, Rathke cleft cysts, optic nerve gliomas and craniopharyngiomas. Conventional sellar floor reconstruction techniques are inadequate for the defects that result from these extended approaches due to larger dural and bone defects [7]. Because the arachnoid membrane is intentionally opened, the continuous higher flow leakage of CSF requires closure techniques that can withstand increased CSF pressure. In contrast to repair of the sella, it is often difficult to detach the dura from the skull base in extended approaches to the planum and tuberculum, preventing watertight extradural closure. This is due to the close proximity of the two opticocarotid recesses medially and by the parasellar carotid protuberances laterally. Factors such as the area and volume of skull base resection, extent of the communication between the cranial cavity and the nasal cavity, adequacy and geometry of the remaining dural and bone margins, previous intranasal surgery, and the possibility of increased CSF pressure should be considered in designing the skull base repair.

Different reconstructive strategies continue to evolve for reconstructing planum and tuberculum defects depending on the dimension and location of the defects. These reconstructive strategies utilize a multilayer combination of bone and dural substitutes placed in an inlay (i.e. underlay) and onlay (i.e. overlay) fashion in the intradural, extradural (i.e. epidural), and extracranial spaces. Cavallo et al. [7] in a series of 21 patients who underwent the extended endoscopic transsphenoidal approach for suprasellar lesion excision describe different techniques for repairing these defects. In the extradural (overlay) technique, a large piece of lyophilized human pericardium (Tutoplast) exceeding the size of the skull base defect is placed in the extradural space as an onlay graft. A piece of a co-polymer of L-lactic acid and glycolic acid (LactoSorb) is wedged above the Tutoplast (also in the extradural, intracranial space), holding the dural substitute in the overlay position. In the intradural (inlay) technique, a piece of Tutoplast cut slightly larger than the skull base defect is placed in the intradural space. A piece of LactoSorb is wedged in the intradural space, superficial to the Tutoplast, to seal the osteodural defect. This technique is used after tuberculum sellae meningiomas in which the size of the dural and bone openings is exactly the same due to the necessity of extensive removal of infiltrated dura. Due to lack of a sufficient bone ledge, LactoSorb cannot be wedged in the extradural space. In the intraextradural (inlay-overlay) technique, a fragment of pericardium is inserted intradurally and a piece of LactoSorb is wedged extradurally, covering the area beneath the osteodural defect. This technique is particularly useful in cases of giant adenomas, suprasellar craniopharyngiomas and optic nerve gliomas. The fully extradural (overlay) technique is the most effective for skull base defects over the planum and tuberculum. The placement of the apposed dural and bone substitutes in the epidural space reduces the risk of migration and material displacement and is most effective at making the closure watertight [7].

Vascularized local and regional tissue flaps have increasingly been utilized to repair planum and tuberculum skull base defects. Vascularized tissue is used to complete the reconstruction, to increase the effectiveness of the reconstruction, and to facilitate wound healing. The most commonly used flap is the Hadad-Bassagasteguy nasoseptal flap, which utilizes the mucoperichondrium and mucoperiosteum of the nasal septum. The flap is supplied by the nasoseptal arteries from the posterior nasal artery, one of the terminal branches of the sphenopalatine artery. The large surface area of the nasoseptal flap allows great versatility of movement. It is capable of reaching any single segment of the ventral skull base, including the sella turcica, planum sphenoidale, clivus or cribriform plate. At its largest dimension, the nasoseptal flap is able to cover an entire anterior craniofacial defect from the frontal sinus to the planum sphenoidale and from orbit to orbit [18, 19]. The endoscopic-assisted pericranial flap has also been used for endoscopic reconstruction of the anterior cranial base given its ideal pedicle location. It is an axial flap based on the supraorbital and supratrochlear arteries, which yields large, durable vascularized tissue that can cover the entire skull base. It is particularly well suited for reconstruction of cribriform and planar defects and can be extended to cover defects of the sella [19, 20].

The Pittsburgh group uses a multilayer closure technique for extended skull base reconstruction. They place an initial subdural inlay graft between the brain and dura mater and favor collagen matrix (Duragen, Integra Life Sciences, Plainsboro, N.J., USA) because it affords the best tissue-handling properties as an inlay graft. The subdural graft should be slightly redundant to extend beyond the dural margins, ideally 5–10 mm in all directions. The pliability of this collagen matrix allows for gentle manipulation around critical neurovascular structures. Duragen is not used alone, because it does not offer sufficient resistance to CSF pressure. CSF will leak through the center of the graft even if the edges have sealed. Intradural fat grafts are avoided due to concern for potentially reproducing the mass effect associated with the initial tumor. In their earlier repairs, they began with an epidural inlay graft (placed between the dura mater and the skull base). However, in their experience they found it difficult to develop an epidural plane, as many skull base defects lack a bone ledge under which to tuck the inlay graft due to the close proximity of neurovascular structures. Planum sphenoidale or the lateral sphenoid wall defects are closely adjacent to the internal carotid artery or optic nerve. An onlay graft is placed epidurally on top of the Duragen inlay graft that extends to overlap the bone margins of the defect. They advocate acellular dermis (AlloDerm, LifeCell, Branchburg, N.J., USA) as their material of choice for the onlay graft due to its rapid epithelialization and ingrowth of fibrous tissue which are valuable in preventing graft migration and early separation. A moderate thickness (0.3–0.7 μm) of acellular dermis, adequately presoaked in saline, is preferred. It is also important that the graft be made to extend beyond the margin of the bone defect. Before placing the graft, all mucosa must be removed from the bone on which the onlay graft is placed to permit the direct contact required for re-vascularization. A single graft without folds is placed as flat as possible to reduce the risk of CSF channel formation. A fat graft is placed obliterating the sphenoid sinus to promote vascularization and healing, applying pressure to and reducing the risk of migration of the grafts underneath. Fibrin glue is then placed over the fat graft and a balloon stent inflated external to the sphenoid sinus to keep the repair in place. The balloon stent (12F Foley catheter) is carefully inflated to avoid compression and injury of any exposed neurovascular structures [5]. Since the development of vascularized local tissue flaps, the choice of overlay graft material has shifted from avascular heterologous grafts to vascularized flap reconstruction.

In their experience, the most common reason for failure of the repair was either the development of a small persistent channel of CSF at the most dependent site of the repair or migration of the graft away from the bone margin of the defect superiorly. CSF leaks varied significantly based on the body habitus of the patient, the type of disease being treated, the site of surgery and whether the arachnoid cisterns and ventricles were exposed [5].

Laufer et al. [17] in their series of endoscopic, endonasal extended transsphenoidal, transplanum transtuberculum approaches for resection of suprasellar lesions vary their reconstruction technique depending on the location of the skull base defect. In closing prechiasmal defects, fat is first placed in the tumor cavity to fill the dead space. An epidural inlay graft (between the dura mater and the skull base) is held in place with a piece of bone or firm synthetic material. A watertight onlay of fibrin matrix (Tisseel, Baxter BioSciences, Deerfield, Ill., USA) or polymerized hydrogel (DuraSeal, Confluent Surgical, Waltham, Mass., USA) seals the closure. In closing postchiasmal defects, because the third ventricle is open, packing with fat is avoided to prevent its dislodging and causing possible obstructive hydrocephalus. An inlay graft is held in place with a piece of bone or firm synthetic material. A watertight onlay of DuraSeal seals the closure. The sphenoid sinus is then filled with thrombin-infused gelatin matrix (FloSeal, Baxter USA) to buttress the closure and to aid in hemostasis.

Reconstruction begins with the obliteration of the dead space with fat to prevent pooling of CSF at the bony defect. The exception is patients in whom the third ventricle has been widely opened, as there is increased risk of fat migration and obstructive hydrocephalus. A gasket-seal closure is performed consisting of a soft tissue graft (fascia lata or AlloDerm) placed in the extradural space and centered over the defect with the edges of the graft exceeding the bony defect circumferentially. Next, a piece of rigid material (vomer or Medpore) is fashioned to be the same size as the bony defect. The rigid graft is countersunk into the bony defect, forming a watertight gasket-seal around the rigid graft. A nasoseptal flap is then rotated to cover the defect, and a tissue sealant is used to secure the multilayer graft in place [17].

A lumbar drain is no longer placed routinely at the time of surgery in extended transplanum, transtubercular approaches because it potentially creates negative pressure that may result in separation of the initial subdural inlay graft from the dural edges [5]. In addition, a lumbar drain can promote separation of the brain away from the operative site and can be counterproductive if a graft inlay is used for closure. In the Schwartz series [17], the risk of CSF leakage was not increased without lumbar drain placement. The use of prophylactic antibiotics for the prevention of meningitis is controversial. A dose of a third-generation cephalosporin is recommended at the time of surgery to prevent colonization of the wound and potential transmission of infection to the CSF (video of a suprasellar meningioma excision and skull base reconstruction available online: see online suppl. video 1).

OSM

Reconstruction of the Clivus and Posterior Skull Base

Dural repair in the clivus is often difficult and requires a multilayer reconstruction technique analogous to repair of planum and tuberculum defects with subdural inlay and epidural onlay graft, covered by a vascularized flap. If the defect is large, it can be occluded with abdominal fat and covered with grafts of fascia lata or Duragen (Integra Life Sciences, Plainsboro, N.J., USA) or Duraform (Duraform, Codman, Raynham, Mass., USA). When the vascularized local tissue flap is transposed over the defect, the flap surface must be carefully apposed without dead space to the underlying fascia or bone. It is critical that the vascularized flap be in direct contact with the margins of the defect and any non-vascularized tissue between the margins of the defect and the flap meticulously removed. The flaps are secured in place with fibrin glue and pieces of Gelfoam. Both local and regional vascularized local tissue flaps may be utilized for clival defects and include the nasoseptal flap described earlier, as well as the posterior pedicled inferior turbinate flap (PPITF) and the posterior palatal flap. The PPITF is based on the inferior turbinate artery, a terminal branch of the posterior lateral nasal artery, which arises from the sphenopalatine artery. The PPITF can be applied directly to dura or denuded bone or may be used over a fat graft [21]. The shorter length and limited arc of rotation of the PPITF in comparison with the nasoseptal flap limits its use. The PPITF is best suited for posterior defects of the sella or parasellar and mid-clival areas, as far anterior skull base reconstructions may not be easily reached due to the limited flap size [1]. The posterior palatal flap is based off the descending palatine artery and has been found to yield a 12–18 cm^2 mucoperiosteal flap based on a 3-cm pedicle. The mucoperiosteum of the palate is transposed into the nasal cavity through the limited enlargement of a single greater palatine foramen [22]. Though theoretically useful, the morbidity and potential for nasopharyngeal incompetence have limited the clinical utility of this flap. The temporoparietal fascial flap can be used when no viable endonasal options are available for reconstruction of clival defects. The temporoparietal fascia is a strong fascial layer that is connected to the overlying fibrous septae of subcutaneous tissue. The blood supply to the temporoparietal fascia originates from the superficial temporal artery, one of the terminal branches of the external carotid artery. During the positioning of the vascularized flap, stretching or compression of the vascular pedicle should be avoided. If the size of the harvested flap is insufficient, free mucosal grafts or multiple rotated local flaps can be used to augment the repair. Antibiotic ointment can be placed over the Gelfoam packing and a Foley catheter balloon is inflated to support the repair [23]. In a series of expanded endoscopic endonasal approaches for the treatment of clival chordomas, Dehdashti et al. [24] report using a multilayer reconstruction technique, starting with fascia lata in an inlay fashion, followed by another layer of fascia lata and fat used as an onlay graft. A nasoseptal flap is used to cover the outer fascia. Fat, fibrin glue, gelatin foam, and the balloon stent are placed to hold the reconstruction in place.

Conclusion

The evolution in endoscopic skull base reconstruction has progressed from the use of free tissue grafts alone to the use of free tissue grafts in combination with vascularized flaps. An increasing armamentarium of local and regional tissue flaps have reduced postoperative CSF leak rates to less than 5%. Future refinements in skull base reconstruction techniques will allow better management of patients at high risk for a postoperative CSF leak, especially those who have been previously irradiated and/or require revision surgery.

References

1 Zanation AM, Thorp BD, Parmar P, et al: Reconstructive options for endoscopic skull base surgery. Otolaryngol Clin N Am 2011;44:1201–1222.
2 Kitano M, Taneda M: Subdural patch graft technique for watertight closure of large dural defects in extended transsphenoidal surgery. Neurosurgery 2004;54:653–661.
3 Pietrzak WS, Verstynen ML, Sarver DR: Bioabsorbable fixation devices: status for the craniomaxillofacial surgeon. J Craniofac Surg 1997;8:92–96.
4 Kveton JF, Friedman CD, Costantino PD: Indications for hydroxyapatite cement reconstruction in lateral skull base surgery. Am J Otol 1995;16:465–469.
5 Kassam A, Carrau RL, Snyderman CH: Evolution of reconstructive techniques following endoscopic expanded endonasal approaches. Neurosurg Focus 2005;19:E8.
6 Menovsky T, de Vries J, Bloss HG, et al: Treatment of postoperative subgaleal cerebrospinal fluid fistulas by using fibrin sealant: a technical note. J Neurosurg 1999;90:1143–1145.
7 Cavallo LM, Messina A, Esposito F, et al: Skull base reconstruction in the extended endoscopic transsphenoidal approach for suprasellar lesions. J Neurosurg 2007;107:713–720.
8 Cappabianca P, Cavallo LM, Valente V, et al: Sellar repair with fibrin sealant and collagen fleece after endoscopic endonasal transsphenoidal surgery. Surg Neurol 2004;62:227–233.
9 Cappabianca P, Cavallo LM, Esposito F, et al: Sellar repair in endoscopic endonasal transsphenoidal surgery: result of 170 cases. Neurosurg 2002;51:1365–1372.
10 Kennedy DW, Cohn ES, Papel ID, et al: Transsphenoidal approach to the sella: the Johns Hopkins experience. Laryngoscope 1984;94:1066–1074.
11 Black PM, Zervas NT, Candia GL: Incidence and management of complications of transsphenoidal operation for pituitary adenomas. Neurosurgery 1987;20:920–923.
12 Spaziante R, de Divitiis E, Cappabianca P: Reconstruction of the pituitary fossa in transsphenoidal surgery: an experience of 140 cases. Neurosurgery 1985;17:453–458.
13 Seiler RW, Mariani L: Sellar reconstruction with resorbable vicryl patches, gelatin foam, and fibrin glue in transsphenoidal surgery: a 10-year experience with 376 patients. J Neurosurg 2000;93:772–775.
14 Lorenz RR, Dean RL, Hurley DB, et al: Endoscopic reconstruction of anterior and middle cranial fossa defects using acellular dermal allograft. Laryngoscope 2003;113:496–501.

15 Esposito F, Dusick JR, Fatemi N, et al: Graded repair of cranial base defects and cerebrospinal fluid leaks in transsphenoidal surgery. Neurosurgery 2007;60(suppl 2):295–303.

16 Tabaee A, Anand VK, Brown SM, et al: Algorithm for reconstruction after endoscopic pituitary and skull base surgery. Laryngoscope 2007;117:1133–1137.

17 Laufer I, Anand VK, Schwartz TH: Endoscopic, endonasal extended transsphenoidal, transplanum transtuberculum approach for resection of suprasellar lesions. J Neurosurg 2007;106:400–406.

18 Hadad G, Bassagasteguy L, Carrau RL, et al: A novel reconstructive technique after endoscopic expanded endonasal approaches. Vascular pedicle nasoseptal flap. Laryngoscope 2006;116:1882–1886.

19 Bhatki AM, Pant H, Snyderman CH, et al: Reconstruction of the cranial base following endonasal skull base surgery: regional tissue flaps. Operative Tech Otolaryngol 2010;21:83–90.

20 Zanation AM, Snyderman CH, Carrau RL, et al: Minimally invasive endoscopic pericranial flap: a new method for endonasal skull base reconstruction. Laryngoscope 2009;119:13–18.

21 Fortes FS, Carrau RL, Snyderman CH, et al: The posterior pedicle inferior turbinate flap: a new vascularized flap for skull base reconstruction. Laryngoscope 2007;117:1329–1332.

22 Oliver CL, Hackman TG, Carrau RL, et al: Palatal flap modifications allow pedicled reconstruction of the skull base. Laryngoscope 2008;118:2102–2106.

23 Fortes FS, Carrau RL, Snyderman CH, et al: Transpterygoid transposition of a temporoparietal fascia flap: a new method for skull base reconstruction after endoscopic expanded endonasal approaches. Laryngoscope 2007;117:970–976.

24 Dehdashti AR, Karabatsou K, Ganna A, et al: Expanded endoscopic endonasal approach for treatment of clival chordomas: early results in 12 patients. Neurosurgery 2008;63:299–309.

James N. Palmer
Department of Otorhinolaryngology, Head and Neck Surgery
University of Pennsylvania School of Medicine, Hospital of the University of Pennsylvania
Ravdin Building 5th Floor, 3400 Spruce Street
Philadelphia, PA 19104 (USA)
E-Mail james.palmer@uphs.upenn.edu

Bleier BS (ed): Comprehensive Techniques in CSF Leak Repair and Skull Base Reconstruction.
Adv Otorhinolaryngol. Basel, Karger, 2013, vol 74, pp 130–137 (DOI: 10.1159/000342289)

Quality Control Approach to Cerebrospinal Fluid Leaks

Carl H. Snyderman[a,b] · Paul A. Gardner[b]

Departments of [a]Otolaryngology and [b]Neurological Surgery, University of Pittsburgh School of Medicine, Pittsburgh, Pa., USA

Abstract

Cerebrospinal fluid (CSF) leaks are one of the most common complications of endoscopic sinus and skull base surgery, and are associated with significant morbidity and cost. A quality improvement program can be designed for individual surgeons or groups of surgeons to monitor outcomes and implement changes to decrease the risk of a postoperative CSF leak. Analytic tools include a root cause analysis, statistical analysis of risk factors, and predictive modeling. Monitoring of outcomes is accomplished with a time series plot or run chart. A circular action loop for assessing and implementing changes facilitates incremental progress and creates a learning culture for the institution.

Quality of patient care is important to everyone involved in health care including patients, physicians, hospital administrators, insurance providers, government, consumer groups, and businesses. Societal forces continue to exert pressure on the United States health care system to improve quality while reducing costs. As individual surgeons and members of the medical profession, we have moral and practical obligations to work toward these goals as well.

Cerebrospinal fluid (CSF) leaks are a quality issue. Every iatrogenic or postoperative CSF leak can be considered an error in patient management with the potential for additional harm (meningitis), intervention (surgical procedure), or cost (additional care, delayed discharge). The goals of management are identification of risk factors for CSF leaks with individualization of care, monitoring of outcomes to detect changes in CSF leak rates, standardization of processes to minimize risk of adverse outcomes, and implementing new processes with monitoring of the impact of quality improvement programs. This chapter will describe how to design and implement a quality control program relative to CSF leaks but has obvious broad application to all aspects of surgical care.

Table 1. Steps of a root cause analysis

Step	Process	Activity
1	identify the problem	who, what, when, where
2	list major factors	systems, equipment, materials, external forces, and people
3	identify possible causes	fishbone diagram
4	analyze the diagram	investigate; action plan

Identification of Risk Factors

A common frailty of human cognition is to selectively remember bad outcomes and more recent events, resulting in an anecdotal approach to patient care. Systematic review of all cases is necessary to minimize cognitive bias. Statistical techniques can be employed to identify risk factors for CSF leaks and stratify patient populations into risk groups. The results of such analyses are often counterintuitive and may provide novel insights that result in new techniques and better patient outcomes. With this knowledge, different treatment strategies can be implemented for different risk groups.

The simplest technique is to routinely examine adverse outcomes and perform a root cause or cause and effect analysis. A root cause analysis is a systematic way of identifying contributing factors to an outcome (table 1). Step 1 is to identify the problem in a succinct manner (e.g. postoperative CSF leak). Step 2 is to identify the major factors that may be part of the problem. Some general categories that are often used in industry but are applicable to health care include systems, equipment, materials, external forces, and people. Step 3 is to identify possible causes within each of these categories. It is helpful to draw a 'fishbone' diagram that links all of the possible causes. An example of a root cause analysis for CSF leak is shown in figure 1. Each cause can be broken down further into possible subcauses. Step 4 is to analyze the diagram and identify causes that need further investigation or are best suited to intervention.

Identification of significant risk factors in a patient population can be done prospectively or retrospectively. Prospective analysis requires a priori identification of factors that are considered to be potentially significant. Potential risk factors for a postoperative CSF leak can be categorized: patient factors, surgical factors, postoperative factors, etc. A list of potential risk factors is provided in table 2. Advantages of prospective analysis are inclusion of most factors that are likely to be relevant and better quality of data collection. Disadvantages include increased work of data collection and potential inclusion of irrelevant factors. Retrospective analysis of data is performed post hoc and attempts to identify risk factors after the outcomes

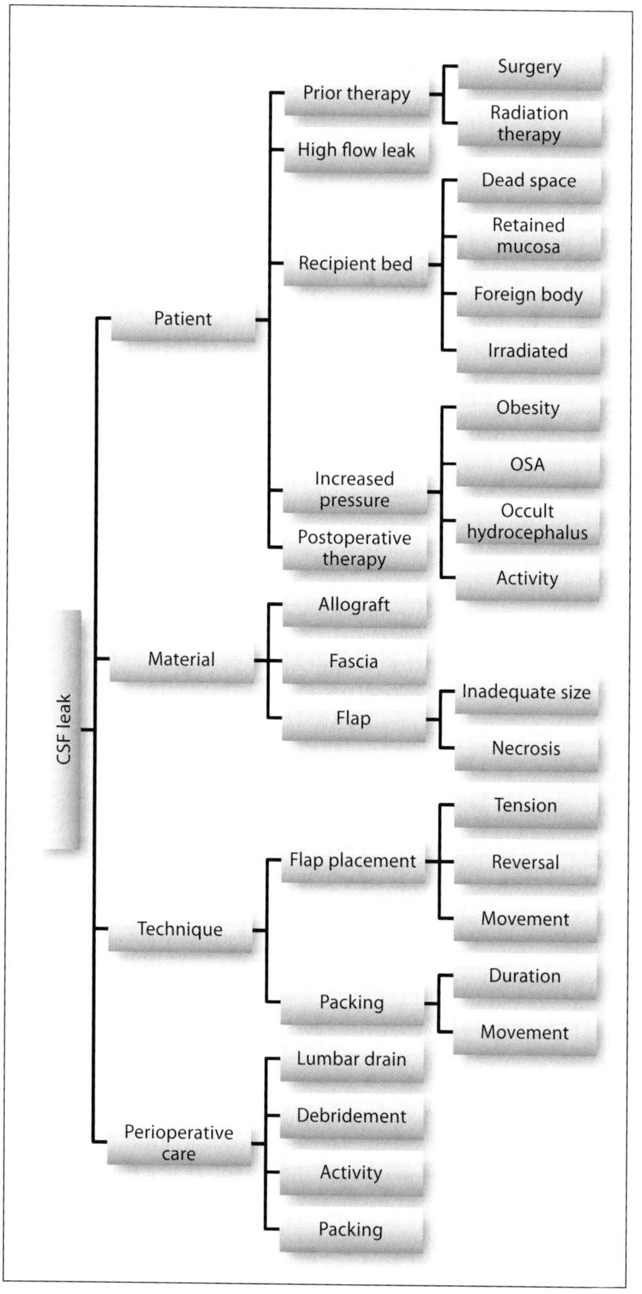

Fig. 1. Hypothetical cause and effect diagram for a patient with a recurrent craniopharyngioma who develops a postoperative CSF leak.

are known. The quality of data collection is often impaired due to limited access to records, poor documentation and missing data. Despite its limitations, retrospective data analysis has the potential to identify important causal relationships and provides access to a larger patient population. There are multiple publications looking at potential risk factors for postoperative CSF leaks [1–5]. Hegazy et al. [1]

Table 2. Potential risk factors for cerebrospinal fluid leak

Patient factors	Material	Technique	Perioperative care
Prior therapy	allograft	flap harvest	lumbar drain/CSF diversion
High flow leak	nonvascularized autograft	inlay graft	debridement
Recipient bed/defect size	vascularized flap	flap placement	patient activity/positioning
Increased CSF pressure	microvascular free flap	packing	packing
Tumor type		tissue glue	
Nutritional status		length of surgery	
BMI			

performed a meta-analysis of the literature and demonstrated that multiple surgical techniques and reconstructive materials were effective. In a follow-up study, Zweig et al. [2] demonstrated that elevated CSF pressure (hydrocephalus) was associated failure of repair. In a study of 70 consecutive high-flow leaks reconstructed with a nasoseptal flap, the CSF leak rate was 5.7% [4]. Risk factors in this study group were pediatric patients, large dural defects, and radiation therapy. In a systematic review of published evidence regarding endoscopic skull base reconstruction of large dural defects, the overall CSF leak rate was 11.5% (70/609): 15.6% leak rate (51/326) for free grafts and a 6.7% leak rate (19/283) for vascularized reconstructions ($p = 0.001$) [5].

Statistical analysis of the data is designed to identify individual risk factors for the outcome and assess their interdependence. A univariate analysis is first performed to identify the most important individual risk factors. A multivariate analysis of this subset of risk factors then examines the interdependence of these factors to determine the relative contribution of each to the outcome. For example, a retrospective analysis of 1,000 sequential endoscopic endonasal surgeries of the skull base was performed to determine the incidence of infectious complications and contributing factors. Of 15 potential risk factors, 10 of them were significant with univariate analysis, but only 4 achieved significance with a logistic regression model (table 3) due to covariance of some factors.

Logistic regression allows one to build a predictive model of the outcome:

- Logit $p = \ln[p/(1-p)] = \beta 0 + \beta 1X1 + \beta 2X2 + .. + \beta pXp$
- p is the probability that the event Y occurs, $p\ (Y = 1)$
- $p/(1-p)$ is the 'odds ratio'
- $\ln[p/(1-p)]$ is the log odds ratio, or 'logit'
- $\beta 0$: intercept

Table 3. Multivariate analysis of risk factors for meningitis (data extracted from prior study [6])

Analysis	Significant risk factors
Univariate	male gender ($p = 0.03$), prior history of craniotomy ($p = 0.0004$), duration of surgery ($p = 0.01$), complexity level of surgery ($p = 0.003$), lumbar drain ($p=0.004$), early reoperation ($p = 0.005$), EVD or VP shunt preoperatively ($p = 0.01$), EVD or VP shunt postop ($p = 0.0003$), postoperative CSF leak ($p < 0.001$), endoscopic repair CSF leak ($p = 0.04$)
Logistic regression (multivariate)	male gender ($p = 0.03$), prior history of craniotomy ($p = 0.002$), complexity level of surgery ($p = 0.003$), postoperative CSF leak ($p < 0.001$)

- $\beta 1$–βp: explanatory variables (represent the fraction by which the outcome/logit is impacted by a unit change in X)
- The coefficients $\beta 1$, $\beta 2$, βp are such that the sums of the squared distance between the observed and predicted values (i.e. regression line) are smallest.

Using the previous analysis of infectious complications, the predictive model is expressed:

- $p_{meningitis} = e^f/(1+e^f)$ where
- $f = -6.67268 + 1.08733(\text{male gender}) + 1.15009(\text{prior surgery}) + 2.03158(\text{complexity level}) + 1.73684(\text{presence of EVD}) + 1.54911(\text{postoperative CSF leak})$.

The variables are either absent (0) or present (1), so that absent conditions disappear from the equation. This model is specific to the patient population studied and may not have applicability to other patient populations. It does allow the surgeon to categorize patients into different risk groups and manage them accordingly. For example, a high-risk patient (male gender, prior surgery, presence of EVD preoperatively, and needing a high complexity surgery) may have a different reconstructive technique or postoperative management to minimize the risk of a CSF leak.

Monitoring of Outcomes

Cerebrospinal fluid leaks are one of the most frequent complications of skull base surgery and are a useful metric for monitoring the success of a skull base surgery program. Surgeons have a responsibility to monitor their own results in order to benchmark their outcomes relative to best practices or standard of care. Continuous monitoring of outcomes also reveals significant deviations from baseline that merit further investigation into cause for corrective action or indicates success of interventions.

The primary goal of quality improvement activities is reduction of variance. Statistical processes are designed to monitor process variation, diagnose causes of

Fig. 2. Time series plot of CSF leaks in 1,000 consecutive cases of endoscopic endonasal surgeries (data extracted from prior study [6]).

variation, and eliminate causes of variation. Quality control techniques that are regularly employed in industry to detect errors and minimize variations in production processes can be applied to surgical outcomes. The most common of these is a time series plot or run chart. Typically, the outcome of interest is plotted by order of surgical procedure or patient. Since the outcome of interest here (CSF leak) is a yes (1) or no (0), positive events are designated by a deviation from baseline (fig. 2). A graph provides a rapid method to assess changes over time. Clustering of cases by volume or time period allows comparison. For example, the data in figure 2 can be analyzed in groups of 100 patients to determine if there are significant differences in the rate of postoperative CSF leaks (table 4). Although there is no significant difference between sequential groups, there appears to be a downward trend in the data. In fact, a comparison of the first half of the series (88/500) to the second half of the series (53/500) reveals a significant decrease in CSF leaks ($p = 0.002$; χ^2 test). This is more apparent when displayed as a graph (fig. 3). Changes in CSF leak rates should prompt an exploration of possible causes. The increase in CSF leaks in the 3rd cohort could reflect a change in the patient population (more complex surgeries) or a change in technique. Continued improvement in subsequent cohorts may reflect a learning curve of the surgeons or incremental improvements in surgical technique. For rare events, individual occurrences can simply be noted as they occur. For example, outcomes could be expressed as one CSF leak in the last quarter or 90 days since the last CSF leak. Time intervals for comparison should be long enough to have a suitable denominator for expression of a ratio but short enough to be practical.

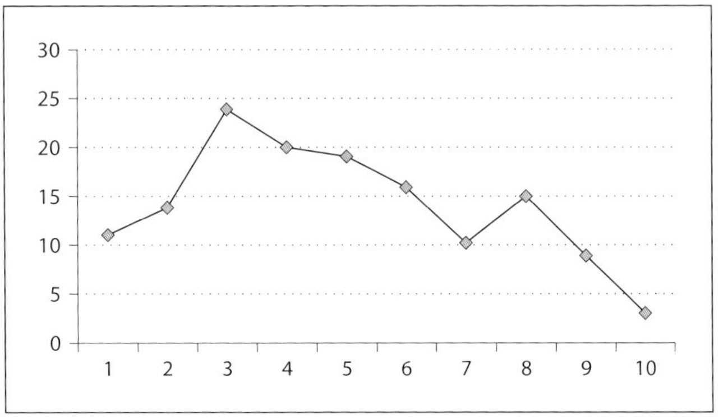

Fig. 3. CSF leaks in sequential groups of 100 patients undergoing endoscopic endonasal surgery (data extracted from prior study [6]).

Table 4. Subgroup analysis of time series plot of CSF leaks in 1,000 consecutive cases of endoscopic endonasal surgeries (data extracted from prior study [6])

Group	Number of patients	CSF leak rate, %	Significance (p value)
1	100	11	–
2	100	14	0.67
3	100	24	0.10
4	100	20	0.62
5	100	19	1.00
6	100	16	0.71
7	100	10	0.29
8	100	15	0.39
9	100	9	0.28
10	100	3	0.13

Statistical analysis: Fisher's exact test (two-tailed).

Improving Outcomes

Once the major risk factors for CSF leak have been identified and a monitoring system is in place, it is possible to institute changes to improve the outcome. As noted in table 2, this may involve better risk stratification of patients based on their characteristics, use of different reconstructive materials, modification of surgical technique, and altering postoperative management. Unlike an industrial process where individual changes can be made in a controlled setting, multiple changes in patient care may occur simultaneously in a seemingly haphazard fashion. Where possible, instituting a single change and assessing the CSF leak rate over a suitable time frame will allow assessment of the impact of the change. When multiple changes are occurring continuously, periodic reanalysis of the patient population for risk factors may be necessary.

An iterative feedback loop of analyzing results (plan), instituting another change (do), assessing outcome (study), and consider additional changes based on results (act) results in a continuous improvement process. When applied across all aspects of patient care, this creates a learning culture where errors or suboptimal results are continuously identified and health care providers learn from mistakes. Sharing of information within the institution, and learning from the best practices of others (direct communication or review of medical literature) are important aspects of this process. For rare events, it may be necessary to collaborate with other centers to generate sufficient volume of cases. When a change in a process that involves multiple health care providers is required, it is necessary to effectively communicate the purpose for the change, the process change, and demonstrate the payoff for everyone.

Conclusions

Using the tools described here, every surgeon can implement a quality control program for the prevention of CSF leaks. These tools can be applied to all aspects of patient care and help create a learning environment where patient outcomes are used to improve the care of patients and decrease the costs of health care. Sustainable change requires that everyone in the organization is committed to learning from their mistakes and is continuously looking for opportunities to improve processes that affect quality. Information is shared freely and incremental changes are trialed in a repetitive cycle of innovation.

References

1 Hegazy HM, Carrau RL, Snyderman CH, Kassam A, Zweig J: Transnasal endoscopic repair of cerebrospinal fluid rhinorrhea: a meta-analysis. Laryngoscope 2000;110:1166–1172.

2 Zweig JL, Carrau RL, Celin SE, Schaitkin BM, Pollice PA, Snyderman CH, Kassam A, Hegazy H: Endoscopic repair of cerebrospinal fluid leaks to the sinonasal tract: predictors of success. Otolaryngol Head Neck Surg 2000;123:195–201.

3 Carrau RL, Snyderman CH, Kassam AB: The management of cerebrospinal fluid leaks in patients at risk for high-pressure hydrocephalus. Laryngoscope 2005;115:205–212.

4 Zanation AM, Carrau RL, Snyderman CH, Germanwala AV, Gardner PA, Prevedello DM, Kassam AB: Nasoseptal flap reconstruction of high flow intraoperative cerebral spinal fluid leaks during endoscopic skull base surgery. Am J Rhinol Allergy 2009;23:518–521.

5 Harvey RJ, Parmar P, Sacks R, Zanation AM: Endoscopic skull base reconstruction of large dural defects: a systematic review of published evidence. Laryngoscope 2012;122:452–459.

6 Kono Y, Prevedello DM, Snyderman CH, Gardner PA, Kassam AB, Carrau RL, Byers KE: One thousand endoscopic skull base surgical procedures demystifying the infection potential: incidence and description of postoperative meningitis and brain abscesses. Infect Control Hosp Epidemiol 2011;32:77–83.

Carl H. Snyderman, MD, MBA
UPMC Center for Cranial Base Surgery
200 Lothrop Street, EEI Suite 500
Pittsburgh PA 15213 (USA)
E-Mail snydermanch@upmc.edu

Bleier BS (ed): Comprehensive Techniques in CSF Leak Repair and Skull Base Reconstruction.
Adv Otorhinolaryngol. Basel, Karger, 2013, vol 74, pp 138–147 (DOI: 10.1159/000342290)

Postoperative Care following Skull Base Reconstruction

Vijay R. Ramakrishnan[a] · Allen Waziri[b]

Departments of [a]Otolaryngology and [b]Neurosurgery, University of Colorado, Aurora, Colo., USA

Abstract

Postoperative care following skull base surgery and reconstruction is critical for achieving optimal outcomes and patient satisfaction. The choice of reconstructive technique and grafting materials may significantly influence the postoperative planning and monitoring for each patient. The overarching goals of postoperative management include avoidance and early recognition of complications, facilitation of wound healing at the surgical site(s), and quick recovery for the patient. Planning for postoperative care is initiated in the preoperative and intraoperative settings. Throughout the reconstruction, the surgical team must actively and meticulously evaluate the potential effects of each maneuver in regard to the subsequent clinical course. During the immediate postoperative period, it is helpful to have clinical protocols in place that address potential surgical complications, allow for early detection of neurological or hormonal dysfunction, and maximize patient safety and comfort. Longer-term postoperative follow-up continues until symptoms resolve and the endoscopic examination is normal. During this time period, well-defined goals and issues of concern should be clearly provided to the patient. In addition, disease-dependent schedules of surveillance are critical for patients with potentially recurrent pathology. This chapter will address decision-making for such topics during postoperative care following skull base reconstruction.

Intraoperative Planning for Postoperative Care

Spinal Drainage

The need for CSF diversion after skull base procedures has been contested. Routine use of a lumbar drain after endoscopic skull base surgery may not be required [1], although a more significant role appears in large skull base defects, in anatomic locations with high CSF flow, for patients with obesity or other mechanisms of intracranial hypertension, or when the surgeon is less confident about the stability of the reconstruction. However, active postoperative use of a lumbar drain does carry some risk. In a recent retrospective review of 65 endoscopic skull base cases in which lumbar drains were used postoperatively, direct drain-related complications occurred in 12% of patients [2].

Overall, in cases where a spinal drain is utilized, we find it helpful to generate an early plan for the nature, extent, and length of time of CSF diversion. Our preference is to perform lumbar drain placement under anesthesia at the onset of the case. This approach is much easier for the surgeon due to maximization of the volume of CSF on insertion (as opposed to placement following significant drainage of CSF during a prolonged intradural procedure). Potentially significant discomfort associated with drain insertion in an unanesthetized patient is also avoided. In addition, the ability to drain CSF in a controlled fashion during the surgical procedure can allow for potentially easier and safer dural incision secondary to brain relaxation.

In general, the presence of a spinal drain does result in decreased mobility of the patient. The use of intermittent rather than continuous drainage may allow patients to ambulate during periods of time when the catheter is clamped; however, risks of disconnection or malfunction are increased with this approach. We generally utilize a drainage protocol that removes 10 ml of CSF every 2 h, which may be titrated upwards or downwards based on symptomatic 'overdrainage' or suspicion for CSF leak. In cases where continuous drainage is utilized, relative immobility may result in subsequent delayed time to ambulation and increased risk of venous thromboembolism. Most surgeons would advocate for continued antibiotic therapy while a spinal drain is in place, although there is no consensus on the value of serial CSF sampling to monitor for potential infection. There are no good data regarding the length of time that a drain should be utilized, but discussions about these issues may be driven intraoperatively by the extent of bony and dural resection, the quality and availability of biological tissue flaps, and patient-specific factors related to successful graft healing (such as prior radiation, medical comorbidities, etc.).

Intranasal Packing and Dressing

An additional issue of importance is the selection and use of intranasal packing and dressing materials that are placed at the end of the surgical procedure. These materials can often improve postoperative healing and stabilization of transposed tissue elements, but a plan for follow-up monitoring and removal of intranasal dressings should be clearly outlined (fig. 1). In addition, use of distant tissue donor sites (fascia lata, abdominal fat, etc.) mandates that these surgical areas be monitored in a typical fashion for signs of infection or poor healing.

Immediate Postoperative Care

General Care

For patients undergoing small, straightforward CSF leak repairs, outpatient management may be considered for a reliable individual with geographic access to

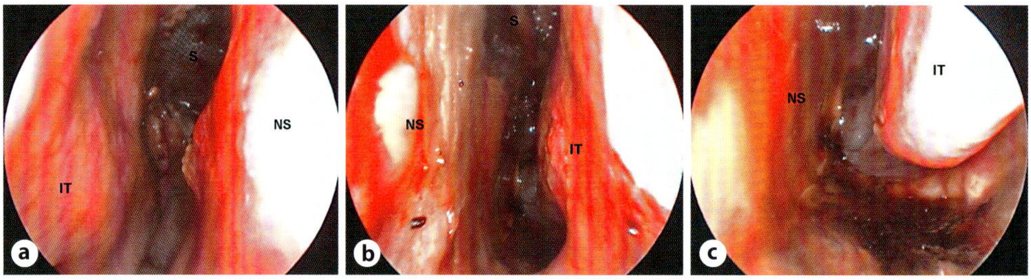

Fig. 1. Examination 2-weeks postoperatively after endoscopic anterior craniofacial resection, reconstruction with left nasoseptal flap. **a** Endoscopy of right nasal cavity shows normal inferior turbinate (IT), healthy nasal septum (NS), and silastic frontal sinus stent (S) superficial to nasoseptal flap. **b**, **c** View of left nasal cavity after removal of septal splint showing nasal septum (NS) and floor with early remucosalization over exposed cartilage.

medical care. However, close follow-up in such cases is mandatory. In our opinion, procedures in which intracranial dissection is performed require that the patient be observed overnight in the hospital. For patients undergoing any level of intradural manipulation our preference is overnight observation in an ICU setting, as the potential for postoperative hemorrhage and rapid neurological decline are complications that cannot be taken lightly. Hourly confirmation of cognitive function, level of alertness, motor activity and preservation of visual fields (particularly in cases involving the midline skull base) may allow for detection of hemorrhage-associated mass affect prior to the onset of permanent neurological injury. Following long cases in which the patient may have a prolonged exit from the effects of anesthesia and delayed neurological recovery, an early postoperative computed tomographic (CT) study will often quickly rule out more sinister surgery-related complications.

ICU-level care is also critical for the hourly assessment of intake and output. These values are particularly important for patients undergoing procedures that involve manipulation of the pituitary gland or infundibulum, in whom diabetes insipidus (DI) may develop in the early postoperative period [3]. Although there are various strategies for the early detection of DI, we utilize a metric that includes urine output (production of greater than 250 ml per hour for 2 consecutive hours) as well as urine specific gravity (<1.005). In most cases, this strategy allows for detection of DI at an early stage, prior to significant and potentially life-threatening elevation of serum sodium.

DVT prophylaxis in patients undergoing skull base surgery remains an area of active discussion. Experience suggests that patients undergoing significant bony removal or tumor resection (particularly meningioma) are at increased risk for the development of DVT. There is no level I evidence providing guidance in regard to timing or specific agent of choice for pharmacological anticoagulation in patients undergoing skull base surgery. In general, we advocate for early ambulation on postoperative day 1, with more aggressive pharmacological intervention reserved for patients who are slow to ambulate or are otherwise neurologically compromised.

Postoperative nasal care is relatively minimal during the period of hospitalization. For local care, saline sprays are used five times daily, or more if patients are willing. Additionally, a face tent providing continuous humidified oxygen may provide some symptomatic relief. The benefits of early initiation of nasal irrigations must be tempered with the potential for confusion of saline effluent with CSF rhinorrhea.

Medications

Postoperative antibiotics with known CSF penetration (vancomycin or cefazolin) are administered intravenously for a minimum of 24 h. Once the patient is able to tolerate oral medications, broad-spectrum oral antibiotics are continued until the nasal packing is removed (amoxicillin-clavulanate). In cases where a spinal drain is left in place, most surgeons would continue the use of prophylactic antibiotics that cover relevant skin flora, although this practice is not supported by the literature.

It could be expected that postoperative pain would be significantly less with endoscopic approaches to the skull base when compared to more traditional open approaches. However, this has not been demonstrated in the literature [4]. Patients are often delayed in their ability to tolerate oral pain medications following skull base surgery, particularly in cases where significant CSF drainage has occurred and nausea may be bothersome. Intravenous narcotic pain medications are therefore useful in the early postoperative period, with a transition to longer-acting oral narcotics in conjunction with acetaminophen once the patient is tolerating oral intake. In general, we avoid the use of nonsteroidal anti-inflammatory medications in the immediate postoperative period due to the hypothetical risk of platelet dysfunction and hemorrhage.

From an endocrine standpoint, there are a number of approaches for monitoring and replacing the pituitary hormones [5, 6]. In all cases in which baseline pituitary function may be compromised, we obtain a comprehensive pre-operative panel of circulating hormone levels (serum cortisol, TSH, free T4, prolactin, LH/FSH, testosterone, and IGF-1). This information often helps guide decision-making in the postoperative period, particularly when considering replacement of corticosteroids. In general, we utilize a rapid (2-day) dexamethasone taper in patients without previously defined pituitary dysfunction undergoing surgical manipulation of the gland or stalk. A cortisol stimulation test may be useful following resolution of the steroid taper to confirm the return of endogenous production. For those patients that develop DI in the postoperative period, management is relatively straightforward with the use of intranasal aqueous pitressin. DI is often a transient complication in these patients, mandating continued monitoring of urine output and serum sodium for titrating ongoing aqueous pitressin treatment. Formal consultation with endocrinology allows for more rarified management of hormonal dysfunction in the immediate as well as long-term postoperative period.

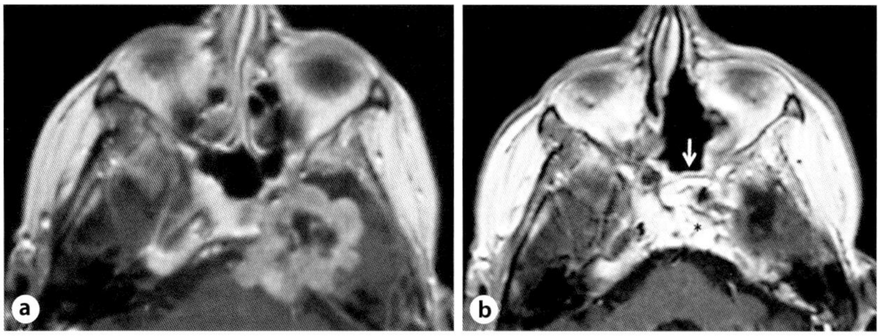

Fig. 2. Pre- and postoperative axial T_1 postcontrast MRI images of left skull base chondrosarcoma. **a** Enhancing tumor with central necrosis involving the clivus, pons, cavernous sinus, and middle fossa. **b** 3 months postoperatively, after endoscopic resection and reconstruction with fat (*) and nasoseptal flap (arrow). No residual tumor is present, although the radiologist's interpretation may not be so definitive.

Radiographic Imaging

Acquisition of postoperative imaging is based on surgeon preference and the nature of the individual pathology. As previously mentioned, noncontrast CT imaging can detect postoperative hemorrhage or significant pneumocephalus that can contribute to neurological decline. In a less acute fashion, early magnetic resonance imaging (MRI) in patients undergoing resection of skull base tumors can provide an important baseline for ongoing monitoring of tumor regrowth or recurrence. We generally advocate for an MRI scan (including intravenous contrast-enhanced images) within 48 h of the surgical procedure, which minimizes the presence of potentially confounding enhancing material that is related to postoperative healing.

Various factors related to skull base repair may render imaging difficult to interpret. Vascularized nasal flaps as well as free fat grafts can mimic pathological tissue on MRI; therefore, surgeon input during the interpretation of such scans is critical (fig. 2).

Early Home Care

Discharge Instructions

On discharge, patients are instructed to restrict certain activities for some period of time. The amount and duration of activity restriction is related to the size of the defect, type of repair, and confidence in the stability of the repair. In general, patients are instructed to avoid heavy lifting (maximum 5–10 lb) and strenuous activity, avoid nose blowing, and sneeze with their mouth open if needed. A regular diet may be resumed at the patient's discretion, but patients are advised to avoid using straws for drinking. Although unlikely to occur, increased pressure within the nasopharynx associated with such activities may allow for transmission of air into the intracranial space, which carries the potential for a 'ball-valve' effect and resulting tension pneumocephalus. Narcotic pain medications are provided, with the expectation that the patient will wean to acetaminophen as soon as reasonably possible. Patients are instructed to take stool softeners as needed to avoid straining, especially if they are concurrently taking narcotic pain medications. The degree of pain is highly variable among patients and may be dependent on individual personality, preoperative use of pain medication, the underlying disease process, and overall extent of surgery.

Prior to discharge, patients are counseled to look for signs and symptoms of postoperative complications. Depending on the extent of surgery performed, the patient and caregivers may need to be aware of CSF leak, meningitis, stroke, orbital hemorrhage, toxic shock syndrome, surgical site infection, or donor site hematoma. In general, major complications would be more likely to occur early during a hospital stay, but early recognition of a complication after discharge could be potentially life-saving.

Management of Epistaxis

A small degree of intermittent epistaxis is expected for up to 2 weeks following surgery. The surgeon may elect to cauterize any raw mucosal edges and/or apply topical hemostatic agents to the surgical cavity at the end of the case in hopes of preventing epistaxis. Mild intermittent nasal bleeding is managed at home with application of topical intranasal decongestant sprays, ice packs, and blood pressure control. However, anything greater than minor bleeding should prompt more formal examination. Nasal packing may not be safe or achievable, depending on the extent of surgery performed and materials left within the nasal cavity. As such, cases of moderate-to-severe epistaxis mandate early and aggressive patient examination. Focal electrocautery or use of hemostatic agents can be attempted within the outpatient clinic setting. If the hemorrhage is anything more than trivial, examination under anesthesia with removal of packing, inspection of the surgical site and reconstruction, and focal application of bipolar cautery or hemostatic agents is preferred. Ligation of the sphenopalatine or anterior ethmoid vasculature can be performed at that time, if desired. In the most extreme cases, angiography and embolization may be indicated if carotid artery hemorrhage is suspected. Of note, embolization may risk necrosis of a pedicled vascularized flap or may be unsuccessful in controlling bleeding, as is seen in up to 25% of cases of posterior epistaxis [7].

Medications

The duration of use of antibiotics after skull base surgery is controversial. Based on more extensive data from open craniofacial surgery study, broad-spectrum antibiotics of variable duration are often recommended, with documented improvements in infection risk, hospital stay, and mortality [8]. Conversely, for endoscopic approaches, perioperative antibiotics with narrow Gram-positive coverage (administered 1 h preoperatively and continued for 24–48 h postoperatively until nasal packing removal) have been advocated as sufficient by Brown et al. [9]. It should be noted that in this study, the majority of cases were pituitary adenomas, the patients received gentamicin nasal spray after systemic antibioticis were discontinued, and an unknown number received antibiotics at follow-up visits for sinusitis. Acknowledging the lack of a definitive evidence-based recommendation, longer courses are frequently given. The practice of routine administration of postoperative antibiotics is followed from sinus surgery performed for inflammatory disease, and is also recommended if nasal packing is left in place [10]. It has not been proven that the use of antibiotics in this manner prevents toxic shock syndrome or meningitis. However, postoperative CSF leak is the circumstance in which CSF rhinorrhea is most associated with bacterial meningitis [11,12].

Medications associated with coagulopathy or platelet dysfunction are restricted for a period of approximately 2 weeks (aspirin, NSAIDs, clopidogrel, herbal supplements, etc.). In endoscopic skull base surgery, unlike functional endoscopic sinus surgery (FESS) for chronic rhinosinusitis, the surgical site may contain large amounts of exposed bone. Re-epithelialization may take some time, and can be associated with pain, thick mucus production, and superficial infection. Saline sprays are routinely used several times daily, and in most cases gentle saline rinses can be of benefit to clear crusts, blood clots, and debris. Tobacco use is restricted for wound healing purposes, as it is known to diminish ciliary function and epithelial regeneration. Anecdotally, perioperative smoking appears to be associated with pain, especially if debridement is performed.

Short- and Long-Term Follow-Up

Early after skull base reconstruction, clinical follow-up and endoscopy plays a crucial role. Packing may be removed along with crusts and associated superficial purulence. Debridement helps restore a comfortable nasal airway for the patient, and may obviate the need for prolonged antibiotics if purulent crusts can be removed. This may also help with patient complaints of nasal dryness, nasal obstruction, epiphora, and malodorous discharge (fig. 3). Nasal crusting and discharge typically improve over the 3- to 6-month time period after surgery, and at 6–12 months sinonasal symptoms are stabilized [13–15].

Early endoscopy allows for visual confirmation that the reconstruction is healing appropriately, and guides decision-making for the timing of return to activity or in planning for the initiation of radiation therapy. The exact follow-up schedule should be guided by the type of procedure and reconstruction performed; a standard paradigm would be packing removal at 1 week, endoscopic examination and debridement at 2–3 weeks, and endoscopic examination at 4–6 weeks, just prior to radiation

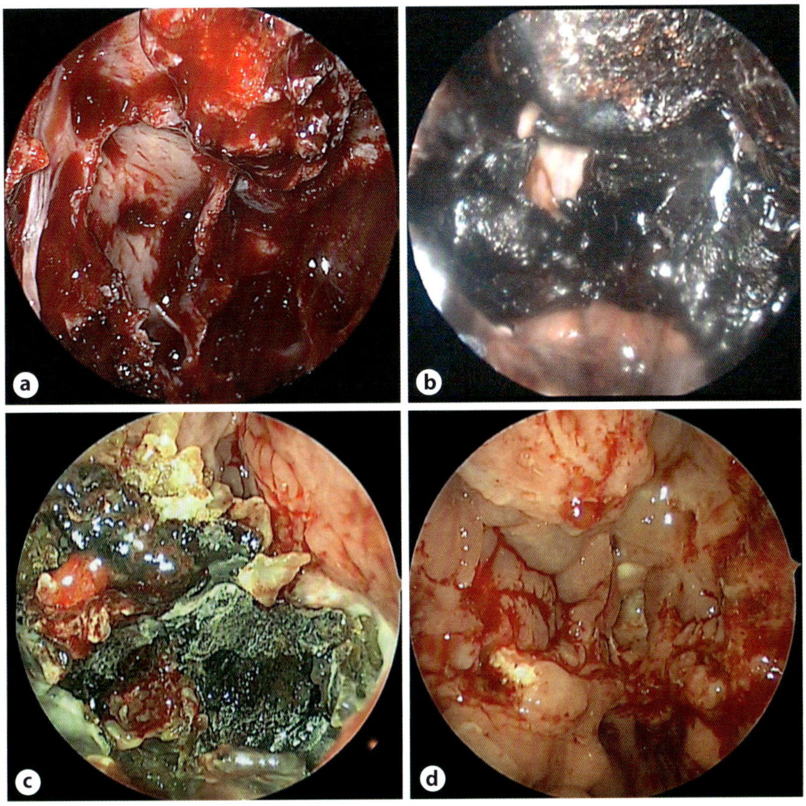

Fig. 3. Endoscopic view of the sphenoid sinus after removal of a low-grade malignancy. **a** Intraoperative view after tumor resection shows the sphenoid intersinus septum, right sphenoid sinus and right optico-carotid recess mucosalized, whereas the left sphenoid and posterior ethmoid regions have exposed dura. **b** Surgicel applied at the end of surgery is still present at 2 weeks postoperatively. **c**, **d** At 6-weeks postoperatively, the patient complained of worsening pain and malodorous drainage. Purulent crusts (**c**) were removed, and edematous underlying mucosa is seen (**d**).

therapy or return to normal activity. During debridement, the area of the graft is not manipulated, and adherent crusts are left intact until they begin to separate spontaneously, usually within 2–6 weeks postoperatively.

Endoscopic examination and long-term clinic follow-up is required in most cases. Aside from surveillance of chronic disease or neoplasm, scarring and mucocele formation may occur in a delayed fashion. Frontal or sphenoid ostia may stenose, particularly after radiation. This occasionally leads to symptoms for the patient, but may also cause confusion for serial radiographic examination (fig. 4). Mucocele formation is rare if a meticulous and focused reconstruction is performed. Although uncommon, placement of a pedicled flap over a defect may obstruct a neighboring sinus and lead to mucocele formation [16]. This is often asymptomatic but can also cause confusion in serial radiographic follow-up of neoplastic disease.

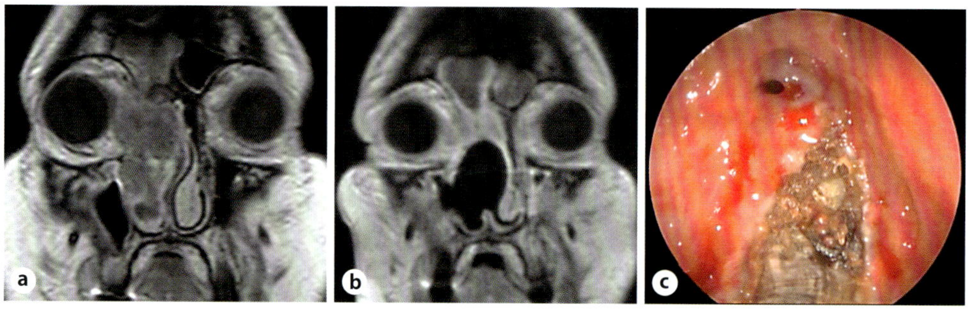

Fig. 4. **a** Coronal T_1-weighted postcontrast MR of a patient with an extensive ethmoid squamous cell carcinoma. **b** Corresponding postoperative image 3 months after orbit-sparing endoscopic-assisted craniofacial resection followed by chemoradiation. Enhancing tissue is seen at the surgical site and within the right frontal sinus, along with fluid in the sinus. **c** Endoscopic examination with a 30° telescope shows stenosis of the frontal ostium as well as partial exposure of the fascia lata graft.

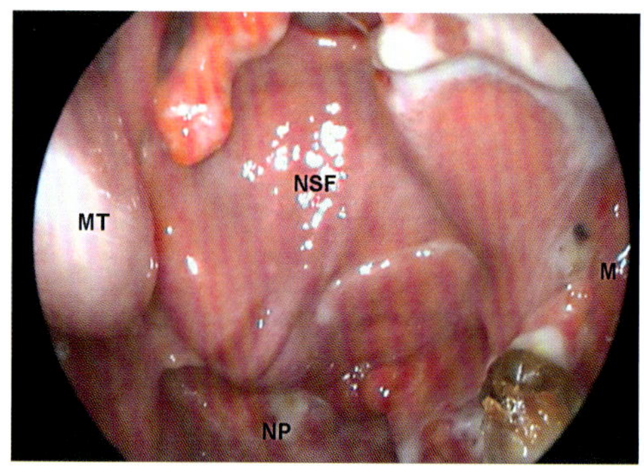

Fig. 5. Endoscopic view of same patient shown in figure 2, approximately 6 weeks postoperatively. The posterior septum has been resected. The nasopharynx (NP), right middle turbinate (MT), and left maxillary sinus (M) are shown for orientation. The nasoseptal flap (NSF) is well healed, and the patient can be allowed to resume normal activity.

Endoscopic confirmation of a well-healed surgical site may reassure both the physician and the patient that it is safe to return to normal activity (fig. 5). For most repairs, patients may return to normal activities with a period of a few weeks to months. In the case of a large skull base defect repaired without a rigid underlay graft, it may be recommended that extreme strenuous activity never be resumed for fear of encephalocele formation although there is no evidence for this, particularly in the era of vascularized flap reconstruction.

Conclusion

Meticulous postoperative care after skull base surgery and reconstruction is absolutely critical for success. Anticipation of the postoperative course begins in the preoperative setting and may influence intraoperative decision-making. Postoperative care in the hospital and outpatient setting is guided by the disease process, extent of surgery performed, size and location of the defect, type of reconstruction utilized, patient-specific factors, and surgeon comfort. Although standard clinical protocols are helpful for many aspects, sufficient literature is not available to guide every decision. In such instances, careful consideration and conservative decision-making will yield optimal results.

References

1 Casiano RR, Jassir D: Endoscopic cerebrospinal fluid rhinorrhea repair: Is a lumbar drain really necessary? Otolaryngol Head Neck Surg 1999; 121:745–750.

2 Ransom ER, Palmer JN, Kennedy DW, Chiu AG: Assessing risk/benefit of lumbar drain use for endoscopic skull-base surgery. Int Forum Allergy Rhinol 2011;1:173–177.

3 Kristof RA, Rother M, Neuloh G, Klingmüller D: Incidence, clinical manifestations, and course of water and electrolyte metabolism disturbances following transsphenoidal pituitary adenoma surgery: a prospective observational study. J Neurosurg 2009;111:555–562.

4 Abergel A, Cavel O, Margalit N, Fliss DM, Gil Z: Comparison of quality of life after transnasal endoscopic vs open skull base tumor resection. Arch Otolaryngol Head Neck Surg 2012;138:142–147.

5 Pereira O, Bevan JS: Preoperative assessment for pituitary surgery. Pituitary 2008;11:347–351.

6 Ausiello JC, Bruce JN, Freda PU: Postoperative assessment of the patient after transsphenoidal pituitary surgery. Pituitary 2008;11:391–401.

7 Klotz DA, Winje MR, Richmon J, Hengerer AS: Surgical management of posterior epistaxis: a changing paradigm. Laryngoscope 2002;112:1577–1582.

8 Kraus DH, Gonen M, Mener D, Brown AE, Bilsky MH, Shah JP: A standardized regimen of antibiotics prevents infectious complications in skull base surgery. Laryngoscope 2005;115:1347–1357.

9 Brown SM, Anand VK, Tabaee A, Schwartz TH: Role of perioperative antibiotics in endoscopic skull base surgery. Laryngoscope 2007;117:1528–1532.

10 Albu S, Lucaciu R: Prophylactic antibiotics in endoscopic sinus surgery: a short follow-up study. Am J Rhinol Allergy 2010;24:306–309.

11 Horowitz G, Fliss DM, Margalit N, Wasserzug O, Gil Z: Association between cerebrospinal fluid leak and meningitis after skull base surgery. Otolaryngol Head Neck Surg 2011;145:689–693.

12 Kono Y, Prevedello DM, Snyderman CH, Gardner PA, Kassam AB, Carrau RL, Byers KE: One thousand endoscopic skull base surgical procedures demystifying the infection potential: incidence and description of postoperative meningitis and brain abscesses. Infect Control Hosp Epidemiol 2011;32: 77–83.

13 de Almeida JR, Snyderman CH, Gardner PA, Carrau RL, Vescan AD: Nasal morbidity following endoscopic skull base surgery: a prospective cohort study. Head Neck 2011;33:547–551.

14 Pant H, Bhakti AM, Snyderman CH, Vescan AD, Carrau RL, Gardner P, Prevedello D, Kassam AB: Quality of life following endonasal skull base surgery. Skull Base 2010;20:35–40.

15 Balaker AE, Bergsneider M, Martin NA, Wang MB: Evolution of sinonasal symptoms following endoscopic anterior skull base surgery. Skull Base 2010;20:245–251.

16 Bleier BS, Wang EW, Vandergrift WA 3rd, Schlosser RJ: Mucocele rate after endoscopic skull base reconstruction using vascularized pedicled flaps. Am J Rhinol Allergy 2011;25:186–187.

Vijay R. Ramakrishnan, MD
Department of Otolaryngology, University of Colorado
12631 E 17th Ave, B205
Aurora, CO 80045 (USA)
E-Mail vijay.ramakrishnan@ucdenver.edu

Bleier BS (ed): Comprehensive Techniques in CSF Leak Repair and Skull Base Reconstruction.
Adv Otorhinolaryngol. Basel, Karger, 2013, vol 74, pp 148–162 (DOI: 10.1159/000342291)

Complications of Skull Base Reconstruction

Mohamad R. Chaaban · Bradford A. Woodworth

Department of Surgery, Division of Otolaryngology, University of Alabama-Birmingham, Birmingham, Ala., USA

Abstract

The skull base is an anatomically challenging location that is unique in that it requires three-dimensional considerations when formulating a reconstruction plan. Various methods have been described for repair including open, endoscopic and endoscopic-assisted approaches. Each of these can have complications that can arise either directly from access to the lesion or defect, or indirectly arising from patient characteristics, anesthetic considerations, as well as postoperative care. It is of utmost importance for the surgeon to select the appropriate technique and surgical access depending on the location of the lesion as well as the patient's specific anatomic variations. In this chapter, we highlight morbidity and mortality arising from various anatomical locations encountered during access for repair of skull base defects as well as those that may arise from lumbar drain placement. We also highlight some treatment protocols from the literature that will aid in the management of these complications. Careful patient selection, review of preoperative imaging, and meticulous technique are of utmost importance to prevent life-threatening problems. Surgeons should have intimate knowledge regarding management as well as methods in place to minimize their occurrence.

Technological and surgical advancements in endoscopic skull base surgery, including powered instrumentation, angled endoscopes, and improved training/experience have led to minimally invasive skull base approaches becoming a standard part of the treatment algorithm for skull base neoplasms [1–4]. At the same time, progress in skull base reconstructive techniques now permit the repair of large breeches in the skull base with low rates of complications. However, skull base defects present a number of unique challenges to successful reconstruction and avoidance of complications due to the inherent three-dimensional structural complexity and proximity to critical neurovascular structures. Despite growing expertise in both open and endoscopic reconstruction in many academic centers, a number of serious and life-threatening complications can occur with rates reported from 16 to 64% [5–7]. Complications may be avoided by careful operative planning and correct selection of the surgical approach to provide the optimal technique for reconstruction. The present chapter will discuss the primary and secondary complications associated with reconstruction of this structurally complex anatomic region and present cutting edge data with fresh perspectives on minimizing morbidity.

Classification and Risk Factors

Complications arising from skull base reconstruction are generally classified as direct and indirect (not arising from the reconstruction itself) or perioperative (early) and late postoperative. Risk factors for the development of complications following skull base reconstruction include both patient-related as well as surgical concerns. Patient-related features include age greater than 60 years, cardiac disease, pulmonary disease, diabetes mellitus, prior surgery, and prior radiation or chemotherapy [8]. Surgical risk factors leading to repair failure and complications include large defect size, presence of high-flow CSF leaks, and underlying intracranial hypertension [9, 10]. In their series of 110 endoscopic repairs, Wise et al. [11] found a number of factors that could potentially contribute to failure including inability to localize the skull base defect, development of new defect, prior sinus or skull base surgery, prior craniotomy, prior radiation therapy, and intracranial infection. In addition, the use of implantable biomaterials such as a Medpor and hydroxyapatite may lead to increased risk of postoperative infectious complications (fig. 1, 2) [12].

Skull Base Anatomy

The skull base acts as the floor of the cranial cavity, separating the brain from the sinuses, nasal cavities, and nasopharynx. There are bony contributions from the ethmoid, sphenoid, occipital, frontal, and parietal bones and three anatomically distinct compartments: the anterior, middle and posterior cranial fossae. Traversing from the nose through the paranasal sinuses and to the skull base are a myriad of nerves, vessels and other structures important to the function of the face and lacrimal systems. While the anatomy of the skull base is described in detail elsewhere, this chapter will discuss complications which are most pertinent to reconstruction according to general anatomic sites [13].

Complications

Scalp

Skull base reconstruction following open craniotomy in traditional or endoscopic-assisted craniofacial resection is primarily accomplished utilizing a pericranial flap. This flap is anteriorly and inferiorly based from the supraorbital and supratrochlear

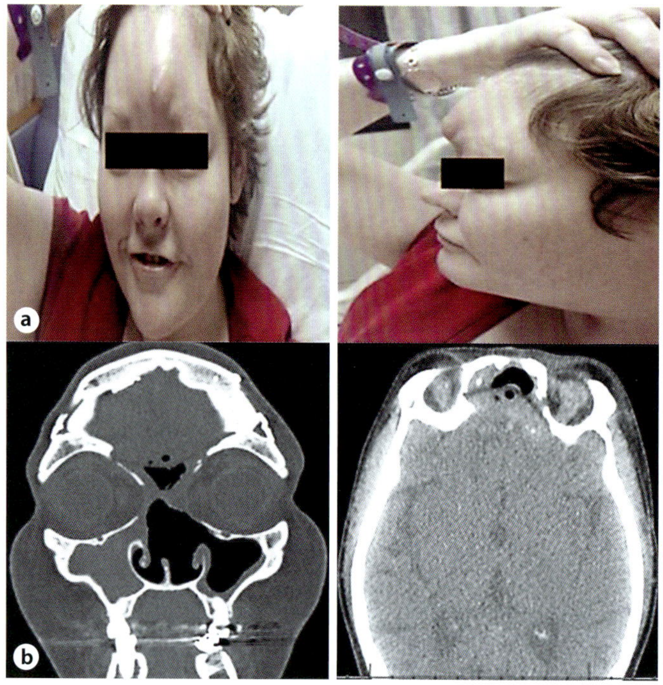

Fig. 1. A patient with a previous craniofacial resection of an adenocarcinoma at an outside institution presented with massive pseudomonas osteomyelitis and purulent drainage from the forehead (**a**). Coronal and axial computed tomography shows osteomyelitis of the frontal bone (**b**).

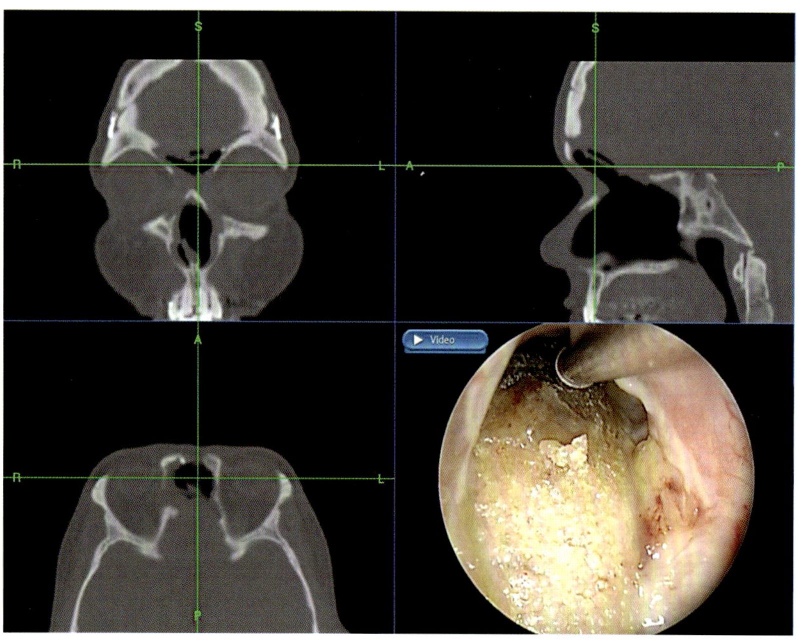

Fig. 2. Patient in figure 1 was found to have chronic mucoid *Pseudomonas aeruginosa* infection from a Medpor implant that was used to reconstruct the skull base as demonstrated in triplanar imaging with endoscopic view. The infected implant and necrotic bone were removed and the anterior frontal space obliterated with a free tissue transfer.

arteries and can extend from the supraorbital rim to the occiput. Failure of the PCF to close the skull base typically occurs at the planum sphenoidale level and has led some surgeons to perform radioanatomic localization and preoperative measurement of scalp incisions to ensure adequate length of the PCF [14]. The technique of endoscopic harvest of the pericranial flap through incisions similar to an endoscopic brow lift was also recently described [15].

Complications from external incisions performed when frontal craniotomy is necessary include alopecia, scar with poor cosmesis, pain, and infection. Alopecia may develop from scalp incisions for harvest of the pericranial flap and is even more predominant in individuals who were previously irradiated [16]. Pain overlying the incision and scalp edema may occur [15]. As with other surgical incisions, infection can occur several days following closure and may require additional antibiotics or drainage in the rare case of abscess development.

Nose

As the gateway to the sinuses and the skull base, the nose is particularly susceptible to deleterious consequences of skull base surgery. Although the majority of complications associated with the nasal airway are not life-threatening, the effect on quality of life can be severe. Significant crusting can develop – particularly in areas of overlying exposed cartilage, bone or synthetic materials. Chemoradiation may result in bone exposure and scarred mucosal surfaces with relentless debris. Resection of large portions of the mucosal surface of the turbinates, particularly the inferior turbinates (if medial maxillectomies are required), can also exacerbate crusting with development of atrophic rhinitis. This condition usually presents with crusting, persistent rhinorrhea, and foul smell or taste as well as paradoxical sensation of nasal airway obstruction [17].

While the nasoseptal flap has become the workhorse for vascularized flap reconstruction in endoscopic skull base surgery, harvest of this flap can lead to disruption of nasal physiology, increased risk of intraoperative bleeding, displacement of the olfactory epithelium, as well as prolonged and persistent nasal cavity crusting postoperatively. Debris as a result of cartilage exposure following a nasoseptal flap can persist for up to 2–3 months. Other vascularized flaps, including the pedicled nasal floor mucosal flap, inferior turbinate/lateral nasal wall flap, and the use of bilateral nasoseptal flaps can create increased pain, dryness, crusting, recurrent epistaxis, and a sensation of nasal congestion [18]. Prolapse or herniation at the skull base following nasoseptal as well as inferior turbinate flap repair has also been reported [19].

Paranasal Sinuses

Endoscopic sinus surgery is naturally required to repair skull base defects with the extent of dissection based upon approach (e.g. transsphenoidal vs. transethmoid), location, etiology, and size of the defect. Like sinus surgery for inflammatory disease, mucosal disruptions, exposed bone, and physical trauma to narrow ostia can lead to iatrogenic sinusitis that requires second procedures [20]. Delayed mucocele formation may develop years later and can complicate surveillance in an area of previously resected neoplasm.

While the maxillary sinus is not in continuity with the base of the skull, this is primarily opened as part of the approach to establish landmarks and prevent the iatrogenic sinusitis from packing and trauma. Primary complications are related to injury to the infraorbital nerve and nasolacrimal duct resulting in numbness of the lip/cheek, and epiphora or incompetent Hasner's valve.

The ethmoid sinus is the primary access point for ethmoid CSF leaks and anterior cranial fossa skull base tumors. During routine CSF leak repair or more extensive endoscopic resection of the anterior cranial base, bleeding may arise from the anterior and posterior ethmoid arteries which lie within the roof of the ethmoid cavity. These arteries require identification and bipolar cauterization if encountered during the repair or resection. The anterior ethmoid artery has no bony covering in up to 66% of individuals [21]. If not controlled, the artery can retract into the orbit and cause orbital hematoma or result in bleeding that impairs the surgeon's visual field. Management of orbital hematoma will be discussed in the orbital section.

The ethmoids also house much of the olfactory epithelium emanating from the cribriform plate. While anosmia is expected in large endoscopic resection of the anterior cranial base, smaller tumors and defects requiring only unilateral approaches can spare sense of smell on the contralateral side and, in some cases, the side of defect closure. However, optimal visualization and adherence to sound oncologic principles should not be sacrificed to spare the sense of smell.

Obstruction of the sphenoid sinus is a frequent problem following skull base repair because surgical approaches to the pituitary gland are so common. In addition, nasoseptal flaps for reconstruction of the anterior skull base may increase sphenoid sinus blockage due to the pedicle's orientation at the sphenopalatine and inferior sphenoid. Careful attention should be paid at the end of the surgical case to ensure sphenoid sinus patency (fig. 3). Bacterial and fungal sinusitis have also been reported following sphenoid sinus blockage during septal flap repair [22].

Major or minor vascular complications may arise from skull base reconstruction within the sphenoid sinus. A dehiscent carotid artery (and optic nerve) is at risk of injury during repair or may be secondarily injured due to instrumentation during the approach to other regions of the skull base. The posterior nasal septal artery may be dissected off the sphenoid face in a submucoperiosteal plane to enable preservation of the artery in the setting of a large sphenoidotomy. Despite this, bleeding from small branches or the main trunk may be encountered and can typically be controlled with cautery. However, delayed bleeding at this site from spontaneous reopening of the vessel secondary to hemodynamic changes (tachycardic and/or marked hypertension) or physical removal of crusting during clinical debridement creates robust epistaxis that may be difficult to control. If this area was cauterized in surgery, crusting

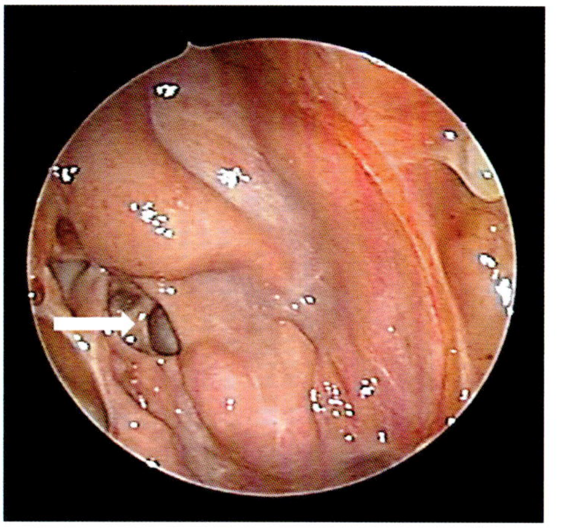

Fig. 3. Postoperative image demonstrating a partially blocked common sphenoid sinus opening after transnasal skull base resection for esthesioneuroblastoma. Fortunately, this was a common cavity where the sinus behind the flap could be visualized through the right open sphenoid (arrow) endoscopically to monitor for closure or other area of tumor recurrence. Thinning the base of the flap should help avoid sphenoid closure.

should be left alone at the base of the sphenoid and sphenopalatine foramen when debrided postoperatively.

The frontal sinus deserves special consideration in terms of complications due to skull base repair. Skull base defects involving the frontal sinus are considered difficult to repair due to anatomic variability, the paucity of tissue available for local transfer, and potential for postoperative stenosis of the frontal recess. Approaches to frontal sinus CSF leaks have traditionally included open techniques such as osteoplastic flap with obliteration or cranialization. The goals of external approaches are to remove sinus mucosa, obliterate dead space, and seal the cranial cavity from the sinonasal tract. Angled (70°) telescopes, the development of advanced frontal sinus instrumentation, and increased expertise have improved endoscopic access to this area for repairing defects without external incisions [23]. While extended endoscopic techniques such as the Draf procedures (Draf IIB and III) are often required for reconstruction access, there is potential for orbital entry and orbital damage due to the use of angled drills [24]. Inadequate coverage with pedicled nasoseptal flaps leading to CSF leak recurrence may also occur. However, some defects up to 3 cm of the posterior table can be reconstructed using this technique [23, 25–27].

Orbit

Orbital injuries have a devastating effect on a patient's quality of life and are a common cause of litigation [28]. Early complications may be vascular, neural, and muscular. Violation of orbital fat causes venous bleeding within the orbital compartment, but ethmoidal arterial injuries are typically the major source for concern as orbital hematoma can develop rapidly into proptosis, elevated ocular pressure, and vascular compromise of the optic nerve and retina. While the incidence of orbital hematoma during or following skull base reconstruction is unknown, its occurrence has been reported [29]. Management of this condition depends on the rate of development, but steps to consider include eye massage, evaluation of eye pressure with tonometer, STAT ophthalmological consultation, medial orbital wall decompression (if intraoperative), lateral canthotomy, administration of mannitol, and intravenous steroids [30].

Optic nerve injuries arise in an indirect (e.g. following orbital hematoma) or direct fashion (e.g. surgical instrumentation). Prevention of this morbid complication is best established by careful review of preoperative CT scans to prepare for anatomic variants such as Onodi cells and dehiscent optic nerves in the sphenoid sinus (present in 4%) [31–33].

Diplopia and enophthalmos are direct consequences of violation of the orbital wall and/or inadvertent injury to the recti muscles – particularly the medial or inferior rectus. Enophthalmos may not be noticeable until after the resolution of periorbital edema. Injury of the medial rectus in FESS is reported in less than 0.5% of the cases [34]. The usual mechanism is penetration through the lamina papyracea and injury to the periorbita using powered instrumentation. Management of such injuries is difficult and the reestablishment of a single monocular field in primary gaze is often the best outcome which can be expected. Repair usually depends on the extent of muscle loss and can include reapproximation, transposition grafts, and more advanced techniques [35, 36].

Pterygopalatine Fossa

A number of neural structures traverse the pterygopalatine fossa (PPF), which should be considered the gateway to middle cranial fossa pathology. Injury to the maxillary nerve (V2) and its associated branches (greater palatine, superior alveolar, and infraorbital) as well as the vidian nerve/sphenopalatine ganglion may cause midfacial numbness, palatal numbness, and dry eye [37]. For access to the PPF, entering through the back wall of the maxillary sinus lateral to the greater palatine nerve and sphenopalatine ganglion and inferior to the infraorbital nerve will leave these structures intact (fig. 4).

The PPF also houses the internal maxillary artery, which typically requires division to gain access to the retropterygoid area. Obtaining endoscopic control of the vessel will be arduous and time-consuming if bleeding occurs from inadequate identification or ligation. Gentle and meticulous separation of pterygopalatine fat from around the IMA will allow definition of the artery for proximal and distal control with the placement of hemostatic clips.

Sellar, Parasellar and Clival Regions

The most fearful complication of sellar and parasellar surgery is a violation of the carotid artery. The internal carotid artery is located in multiple planes as it traverses the posterolateral wall of the sphenoid and cavernous sinus. The vessel is at risk during a transsphenoidal approach to the sella when removing an intersinus septum that careens toward the lateral wall where it inserts at the face of the ascending carotid

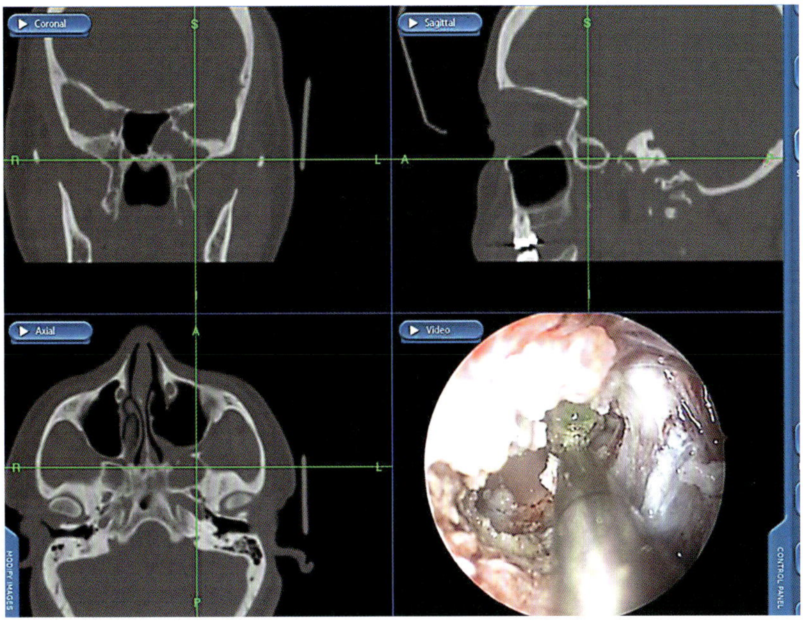

Fig. 4. A transpterygoid approach to the lateral recess of the left sphenoid sinus. The encephalocele is demonstrated at the tip of the suction leaking fluorescein. Because the approach is through the back wall of the maxillary sinus lateral to the greater palatine nerve and inferior to the maxillary nerve, this patient had no anesthesia or dry eye following the procedure.

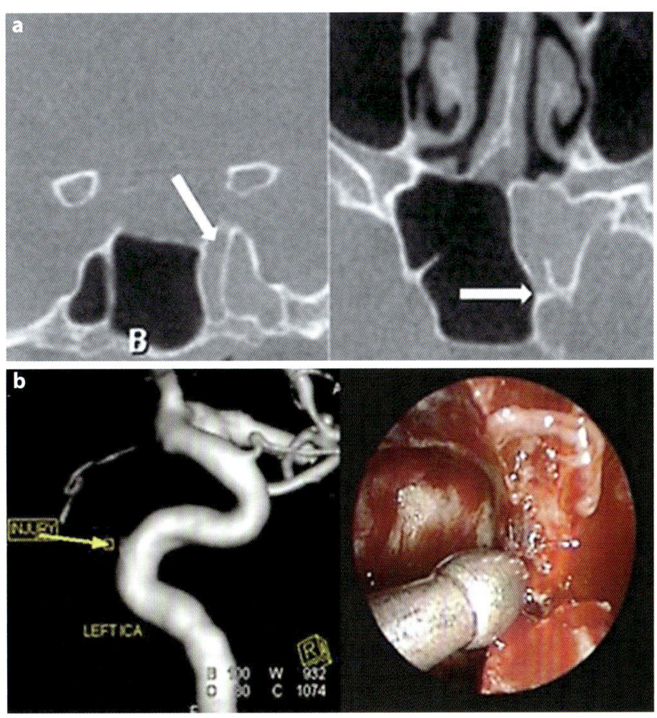

Fig. 5. **a** Coronal and axial CT scan demonstrating an intersinus septum deviating to the ascending portion of the left carotid artery (arrow). Note the anterior location of the artery. **b** During a transsphenoidal approach to a pituitary tumor, the carotid artery is injured during removal of the intersinus septum as noted on the carotid angiogram (yellow arrow – left). Inspection of the area after internal carotid artery coiling by neurointerventional radiology reveals missing bone over the anterior wall of the carotid (right).

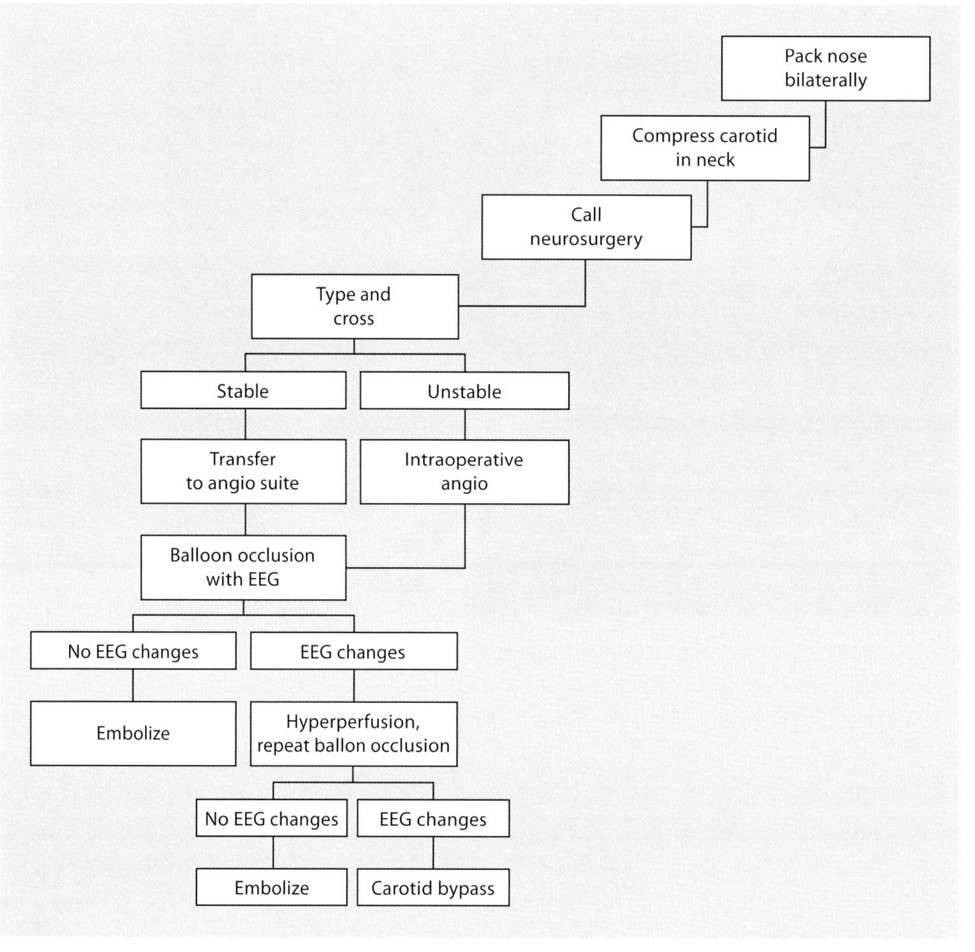

Fig. 6. Algorithm of management of carotid artery injury (The Sofferman protocol). Reprinted with permission from Moeller and Welch [40].

(fig. 5). Carotid artery injury remains one of the most devastating complications of pituitary surgery, and is rarely reported in the literature [38, 39]. Bleeding from the ICA is a life threatening emergency and should be handled in a well-defined fashion similar to the Sofferman protocol (fig. 6) [40].

Impaired visualization from cavernous sinus bleeding in sellar and parasellar surgery can also severely impair visualization and should be controlled prior to proceeding. Cranial neuropathies from instrumenting the cavernous sinus may result in ophthalmoplegia – most commonly the abducens nerve as it runs most medial in the cavernous sinus with an incidence of 1 in 400 in pituitary surgery [41].

Endocrine complications are common following pituitary surgery, but not typically as a direct result of skull base reconstruction. Diabetes insipidus can either be transient or permanent and occurs with an incidence of 0.4–17% after trans-sphenoidal pituitary surgery [42]. However, congenital encephaloceles of the sphenoid sinus can be associated with a bifid pituitary on the lateral walls of an underdeveloped sphenoid and panhypopituitarism is a potential risk during repair [43, 44].

Other Complications

Intrasellar hematoma may transiently worsen vision due to damage to the optic chiasm and/or the optic nerves.

Direct injury to the hypothalamus is one of the major causes of surgical death and usually a result of large tumor resection characterized by significant bleeding. Death has been reported to occur within 1–3 days [45].

Basilar artery injury can arise while drilling for removal of a clival chordoma or chondrosarcoma in addition to C1-C2 endoscopic decompression.

Intracranial

The critical component of a successful endoscopic endonasal reconstruction is a watertight dural seal that provides a complete barrier between the nose and brain. Early complications related to the intracranial space include reconstruction failure with recurrent CSF leak, pneumocephalus, encephalitis, meningitis and/or cerebral abscess [46, 47]. Other intracranial complications include vascular injuries and intracranial bleeding. Knowledge of the cerebral vascular anatomy, such as variations in the origin of the recurrent artery of Heubner (double vessel in 17% and originate mostly close to the anterior communicating artery in 62%) is essential before any skull base approach is attempted [48]. Injury to the anterior cerebral arteries may also occur following resection of a transplanum meningioma or craniopharyngioma.

Other Complications

Tension pneumocephalus develops in up to 5% of open skull base approaches (unknown incidence following endoscopic cranial base surgery) and is thought to arise from ball-valving of the reconstruction material (fig. 7). Some pneumocephalus is expected postoperatively following large endoscopic resection of the skull base, but can be minimized with saline volume replacement [9].

Vascular or neurological insults following reconstruction may lead to seizures.

Biomaterials may lead to infectious complications such as meningitis, encephalitis, and/or the development of abscesses.

Indirect Complications

Systemic complications such as deep vein thrombosis, ulcer rupture, and cardiovascular events arise from anesthesia, prolonged operative time and surgical stress. Complications from lumbar drain utilization include pneumocephalus from decreased pressure gradient, ascending meningitis, arachnoiditis, epidural

Fig. 7. CT scan images of a patient that presented obtunded to the emergency room following 2 weeks following repair of an anterior skull base defect. **a** Coronal CT scan demonstrating the possible site of nasal septal flap dehiscence (white arrow). **b** Sagittal CT scan demonstrating the extent of pneumocephalus.

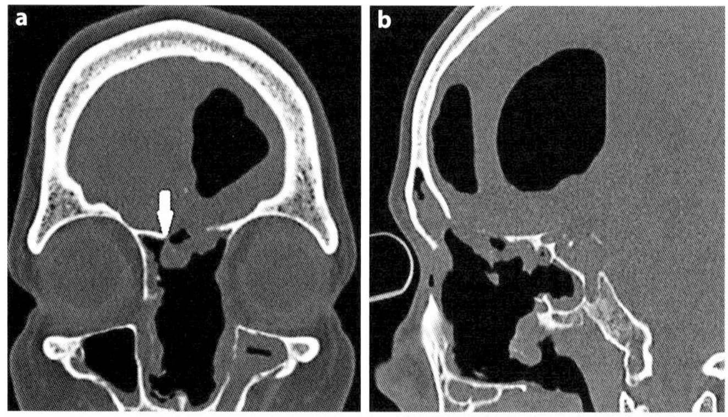

hematoma, as well as more serious complications such as posterior cerebral artery infarction from transtentorial herniation [49, 50]. Minor complications (e.g. headache and nausea) are reported in the literature in up to 59% with a 12.5% incidence of major complications (e.g. pneumocephalus, ascending meningitis, vocal cord paralysis).

Preoperative injection of intrathecal fluorescein through the lumbar drain is useful for localizing defects, identifying multiple CSF leaks, and inspecting for a watertight closure at the conclusion of the case. However, fluorescein is not FDA approved for intrathecal injection because seizures and neurotoxicity have been reported when using higher concentrations or rapid injections [51]. In over 20 years of utilization, a mixture of 0.1 ml of 10% fluorescein diluted in 10 ml of the patient's CSF (or preservative-free sterile saline) slowly injected over 10–15 min has resulted in only one minor complication that could potentially be attributable to its use – an arrhythmia that resolved within 24 h [unpubl. data]. It is still recommended to obtain informed written consent regarding the risks and benefits of intrathecal fluorescein and its lack of FDA approval in all patients.

Minimizing Morbidity and Reconstruction Failure

In addition to careful patient selection and accounting for risk factors, selection of the surgical approach and reconstructive technique is crucial for the prevention of complications. If a craniotomy is required and a pericranial flap is utilized, complete sterilization of the surgical site, use of a beveled incision, and hemostasis with Rainey clips instead of bovie cautery will decrease wound infection and alopecia along the incision [52]. When using the pericranial flap for skull base defects involving the planum sphenoidale, particularly in the setting of previous chemoradiation treatment, a simultaneous closure with an opposing nasoseptal flap provides a double layer of vascularized reconstruction.

Avoiding surgical resection of normal structures (e.g. inferior turbinates) when possible will decrease the risk of crusting and atrophic rhinitis, but should be removed if interfering with tumor extirpation. Saline irrigations with surfactants (e.g. baby shampoo) and topical antibiotics may also be beneficial for crusting [53]. Intranasal synechiae and crusting are managed in the office using topical anesthesia and frequent debridement, if necessary.

Minimizing excessive bleeding for optimal visualization is the key ingredient to avoid intraoperative complications. Topical and infiltrative vasoconstriction and avoidance of mucosal trauma are critical to success. Meticulous bipolar cautery during encephalocele reduction is critical to avoiding retraction of a small dural or cerebral vessel within the sac. The Coblator (ArthroCare ENT, Sunnyvale, Calif., USA) is useful for rapid reduction in the encephalocele sac (both radiofrequency ablation and bipolar) without a concomitant increase in bleeding and also decreases bleeding in sinonasal/skull base tumor removal [54–57]. Avoiding trauma to the anterior ethmoid artery and limiting dissection in the region of the sphenopalatine artery and its branches will also reduce bleeding. The Endoscrub (Medtronic-Xomed Inc., Jacksonville, Fla., USA) device clears the endoscope lens of blood and is particularly helpful for maintaining optimal surgical imaging.

Success rates for endoscopic skull base defect repair are commonly greater than 90% in most series, but decrease with large defect size following tumor extirpation [10, 58–64]. The development and utility of vascularized nasoseptal flaps and other variations of pedicled local reconstructive techniques have been a critical factor in improving success rates for large anterior cranial base defects and high-flow CSF leaks. In fact, a recent meta-analysis regarding endoscopic skull base reconstruction of large cranial base defects demonstrated a decreased failure rate with vascularized reconstruction vs. free grafts (6.7 vs. 15.6%. respectively) [65].

CSF leaks of spontaneous etiology deserve special mention as treatment philosophy has changed over the last 15 years. Although this classification varies in the literature, those leaks that do not have an identifiable cause are generally referred to as 'spontaneous' or 'idiopathic'. In this subset of patients, successful repair rates are dismal with recurrences previously reported anywhere from 25 to 87% over time [66]. These CSF leaks are associated with the highest rate (50–100%) of encephalocele formation, as there are often large meningoencephaloceles herniating through relatively small bony defects [63, 64, 66]. Mounting evidence indicates chronically elevated ICP is the underlying cause of spontaneous CSF rhinorrhea and encephalocele formation in this population. Multiple studies have confirmed the elevated CSF pressure status in these individuals [61, 67]. Findings have affected treatment philosophy where opening lumbar tap pressures and manometry measurements following closure of the defects allow for accurate institution of interventions to decrease intracranial hypertension including acetazolamide and ventriculoperitoneal shunting. Closure rates for spontaneous leaks now approach 95% when intracranial hypertension is adequately controlled [58, 68, 69].

Conclusion

Techniques for skull base reconstruction have rapidly evolved over the last 20 years. Minimally invasive approaches to the cranial base have become a routine part of the treatment algorithm for CSF leaks and skull base defects that result from a number of pathological processes. With growing interest in training programs and the widespread acquisition of skills and experience, there is promise of decreased morbidity and mortality related to skull base surgery in the future.

References

1 Day TA, Beas RA, Schlosser RJ, et al: Management of paranasal sinus malignancy. Curr Treat Options Oncol 2005;6:3–18.

2 Schlosser RJ, Woodworth BA, Gillespie MB, Day TA: Endoscopic resection of sinonasal hemangiomas and hemangiopericytomas. ORL J Otorhinolaryngol Relat Spec 2006;68:69–72.

3 Woodworth BA, Bhargave GA, Palmer JN, et al: Clinical outcomes of endoscopic and endoscopic-assisted resection of inverted papillomas: a 15-year experience. Am J Rhinol 2007;21:591–600.

4 Schuster D, Riley KO, Cure JK, Woodworth BA: Endoscopic resection of intracranial dermoid cysts. J Laryngol Otol 2011;125:423–427.

5 Clayman GL, DeMonte F, Jaffe DM, et al: Outcome and complications of extended cranial-base resection requiring microvascular free-tissue transfer. Arch Otolaryngol Head Neck Surg 1995;121:1253–1257.

6 Kraus DH, Shah JP, Arbit E, Galicich JH, Strong EW: Complications of craniofacial resection for tumors involving the anterior skull base. Head Neck 1994;16:307–312.

7 Urken ML, Catalano PJ, Sen C, Post K, Futran N, Biller HF: Free tissue transfer for skull base reconstruction analysis of complications and a classification scheme for defining skull base defects. Arch Otolaryngol Head Neck Surg 1993;119:1318–1325.

8 Hanasono MM, Silva A, Skoracki RJ, et al: Skull base reconstruction: an updated approach. Plast Reconstr Surg;128:675–686.

9 Blount A, Riley K, Cure J, Woodworth BA: Cerebrospinal fluid volume replacement following large endoscopic anterior cranial base resection. Int Forum Allergy Rhinol 2012;2:217–221.

10 Banks CA, Palmer JN, Chiu AG, O'Malley BW Jr, Woodworth BA, Kennedy DW: Endoscopic closure of CSF rhinorrhea: 193 cases over 21 years. Otolaryngol Head Neck Surg 2009;140:826–833.

11 Wise SK, Harvey RJ, Neal JG, Patel SJ, Frankel BM, Schlosser RJ: Factors contributing to failure in endoscopic skull base defect repair. Am J Rhinol Allergy 2009;23:185–191.

12 Wang B, Wu ST, Li Z, Liu PN: Anterior and middle skull base reconstruction after tumor resection. Chin Med J 2010;123:281–285.

13 Duckert LG: Anatomy of the Skull Base, Temporal Bone, External Ear, and Middle Ear. Otolaryngology: Head and Neck Surgery. San Diego, 1998.

14 Patel MR, Shah RN, Snyderman CH, et al: Pericranial flap for endoscopic anterior skull-base reconstruction: clinical outcomes and radioanatomic analysis of preoperative planning. Neurosurgery 2010;66:506–512.

15 Zanation AM, Snyderman CH, Carrau RL, Kassam AB, Gardner PA, Prevedello DM: Minimally invasive endoscopic pericranial flap: a new method for endonasal skull base reconstruction. Laryngoscope 2009;119:13–18.

16 Fliss DM, Gil Z, Spektor S, et al: Skull base reconstruction after anterior subcranial tumor resection. Neurosurg Focus 2002;12:e10.

17 Dutt SN, Kameswaran M: The aetiology and management of atrophic rhinitis. J Laryngol Otol 2005;119:843–852.

18 Heller JB, Gabbay JS, Trussler A, Heller MM, Bradley JP: Repair of large nasal septal perforations using facial artery musculomucosal (FAMM) flap. Ann Plast Surg 2005;55:456–459.

19 El-Banhawy OA, Halaka AN, Altuwaijri MA, Ayad H, El-Sharnoby MM: Long-term outcome of endonasal endoscopic skull base reconstruction with nasal turbinate graft. Skull Base 2008;18:297–308.

20 Unger F, Haselsberger K, Walch C, Stammberger H, Papaefthymiou G: Combined endoscopic surgery and radiosurgery as treatment modality for olfactory neuroblastoma (esthesioneuroblastoma). Acta Neurochir (Wien) 2005;147:595–601.

21 Araujo Filho BC, Weber R, Pinheiro Neto CD, Lessa MM, Voegels RL, Butugan O: Endoscopic anatomy of the anterior ethmoidal artery: a cadaveric dissection study. Braz J Otorhinolaryngol 2006;72:303–308.

22 Cavallo LM, Messina A, Esposito F, et al: Skull base reconstruction in the extended endoscopic transsphenoidal approach for suprasellar lesions. J Neurosurg 2007;107:713–720.

23 Woodworth BA, Schlosser RJ, Palmer JN: Endoscopic repair of frontal sinus cerebrospinal fluid leaks. J Laryngol Otol 2005;119:709–713.

24 Conger BT Jr, Riley K, Woodworth BA: The Draf III mucosal grafting technique: a prospective study. Otolaryngol Head Neck Surg 2012;146:664–668.

25 Jones V, Virgin F, Riley K, Woodworth BA: Changing paradigms in frontal sinus cerebrospinal fluid leak repair. Int Forum Allergy Rhinol 2012.

26 Virgin F, Baranano CF, Riley K, Woodworth BA: Frontal sinus skull base defect repair using the pedicled nasoseptal flap. Otolaryngol Head Neck Surg 2011.

27 Purkey MT, Woodworth BA, Hahn S, Palmer JN, Chiu AG: Endoscopic repair of supraorbital ethmoid cerebrospinal fluid leaks. ORL J Otorhinolaryngol Relat Spec 2009;71:93–98.

28 Becker SS, Duncavage JA: Malpractice claims in nasal and sinus surgery: a review of 15 cases. Otolaryngol Clin North Am 2010;43:929–932.

29 Dave SP, Bared A, Casiano RR: Surgical outcomes and safety of transnasal endoscopic resection for anterior skull tumors. Otolaryngol Head Neck Surg 2007;136:920–927.

30 Welch KC, Palmer JN: Intraoperative emergencies during endoscopic sinus surgery: CSF leak and orbital hematoma. Otolaryngol Clin North Am 2008;41:581–596.

31 Arslan H, Aydinlioglu A, Bozkurt M, Egeli E: Anatomic variations of the paranasal sinuses: CT examination for endoscopic sinus surgery. Auris Nasus Larynx 1999;26:39–48.

32 Bhatti MT, Stankiewicz JA: Ophthalmic complications of endoscopic sinus surgery. Surv Ophthalmol 2003;48:389–402.

33 Dessi P, Moulin G, Castro F, Chagnaud C, Cannoni M: Protrusion of the optic nerve into the ethmoid and sphenoid sinus: prospective study of 150 CT studies. Neuroradiology 1994;36:515–516.

34 Corey JP, Bumsted R, Panje W, Namon A: Orbital complications in functional endoscopic sinus surgery. Otolaryngol Head Neck Surg 1993;109:814–820.

35 Huang CM, Meyer DR, Patrinely JR, et al: Medial rectus muscle injuries associated with functional endoscopic sinus surgery: characterization and management. Ophthal Plast Reconstr Surg 2003;19:25–37.

36 Penne RB, Flanagan JC, Stefanyszyn MA, Nowinski T: Ocular motility disorders secondary to sinus surgery. Ophthal Plast Reconstr Surg 1993;9:53–61.

37 Hillel AT, Metzinger RC, Nemechek AJ, Nuss DW: Loss of reflex tearing: an expected consequence of juvenile nasopharyngeal angiofibroma. Otolaryngol Head Neck Surg 2005;133:605–610.

38 Frank G, Sciarretta V, Calbucci F, Farneti G, Mazzatenta D, Pasquini E: The endoscopic transnasal transsphenoidal approach for the treatment of cranial base chordomas and chondrosarcomas. Neurosurgery 2006;59:ONS50–57; discussion ONS50–57.

39 Stippler M, Gardner PA, Snyderman CH, Carrau RL, Prevedello DM, Kassam AB: Endoscopic endonasal approach for clival chordomas. Neurosurgery 2009;64:268–277.

40 Moeller CW, Welch KC: Prevention and management of complications in sphenoidotomy. Otolaryngol Clin North Am 2012;43:839–854.

41 Raithatha R, McCoul ED, Woodworth GF, Schwartz TH, Anand VK: Endoscopic endonasal approaches to the cavernous sinus. Int Forum Allergy Rhinol 2012;2:9–15.

42 Laws ER Jr, Kern EB: Complications of trans-sphenoidal surgery. Clin Neurosurg 1976;23:401–416.

43 Woodworth BA, Schlosser RJ, Faust RA, Bolger WE: Evolutions in the management of congenital intranasal skull base defects. Arch Otolaryngol Head Neck Surg 2004;130:1283–1288.

44 Woodworth B, Schlosser RJ: Endoscopic repair of a congenital intranasal encephalocele in a 23-month-old infant. Int J Pediatr Otorhinolaryngol 2005;69:1007–1009.

45 Sethi DS, Leong JL: Endoscopic pituitary surgery. Otolaryngol Clin North Am 2006;39:563–583.

46 El-Banhawy OA, Halaka AN, El-Dien AE, Ayad H: Sellar floor reconstruction with nasal turbinate tissue after endoscopic endonasal transsphenoidal surgery for pituitary adenomas. Minim Invas Neurosurg 2003;46:289–292.

47 El-Banhawy OA, Halaka AN, El-Hafiz Shehab El-Dien A, Ayad H: Subcranial transnasal repair of cerebrospinal fluid rhinorrhea with free autologous grafts by the combined overlay and underlay techniques. Minim Invas Neurosurg 2004;47:197–202.

48 Loukas M, Louis RG Jr, Childs RS: Anatomical examination of the recurrent artery of Heubner. Clin Anat 2006;19:25–31.

49 Ransom ER, Chiu AG: Prevention and management of complications in intracranial endoscopic skull base surgery. Otolaryngol Clin North Am 2010;43: 875–895.
50 Ransom ER, Palmer JN, Kennedy DW, Chiu AG: Assessing risk/benefit of lumbar drain use for endoscopic skull-base surgery. Int Forum Allergy Rhinol 2011;1:173–177.
51 Park KY, Kim YB: A case of myelopathy after intrathecal injection of fluorescein. J Korean Neurosurg Soc 2007;42:492–494.
52 Flowers RS, Ceydeli A: The open coronal approach to forehead rejuvenation. Clin Plast Surg 2008;35: 331–351.
53 Chiu AG, Palmer JN, Woodworth BA, et al: Baby shampoo nasal irrigations for the symptomatic post-functional endoscopic sinus surgery patient. Am J Rhinol 2008;22:34–37.
54 Blount A, Riley KO, Woodworth BA: Juvenile nasopharyngeal angiofibroma. Otolaryngol Clin North Am 2011;44:989–1004, ix.
55 Virgin FW, Bleier BS, Woodworth BA: Evolving materials and techniques for endoscopic sinus surgery. Otolaryngol Clin North Am 2010;43:653–672.
56 Smith N, Riley KO, Woodworth BA: Endoscopic Coblator-assisted management of encephaloceles. Laryngoscope 2010;120:2535–2539.
57 Kostrzewa JP, Sunde J, Riley KO, Woodworth BA: Radiofrequency coblation decreases blood loss during endoscopic sinonasal and skull base tumor removal. ORL J Otorhinolaryngol Relat Spec 2010; 72:38–43.
58 Woodworth BA, Schlosser RJ: Repair of anterior skull base defects and CSF Leaks. Op Tech Otolaryngol 2006;18:111–116.
59 Woodworth BA, Neal JG, Schlosser RJ: Sphenoid sinus cerebrospinal fluid leaks. Op Tech Otolaryngol 2006;17:37–42.
60 Schlosser RJ, Bolger WE: Nasal cerebrospinal fluid leaks: critical review and surgical considerations. Laryngoscope 2004;114:255–265.
61 Mattox DE, Kennedy DW: Endoscopic management of cerebrospinal fluid leaks and cephaloceles. Laryngoscope 1990;100:857–862.
62 McMains KC, Gross CW, Kountakis SE: Endoscopic management of cerebrospinal fluid rhinorrhea. Laryngoscope 2004;114:1833–1837.
63 Gassner HG, Ponikau JU, Sherris DA, Kern EB: CSF rhinorrhea: 95 consecutive surgical cases with long term follow-up at the Mayo Clinic. Am J Rhinol 1999;13:439–447.
64 Schick B, Ibing R, Brors D, Draf W: Long-term study of endonasal duraplasty and review of the literature. Ann Otol Rhinol Laryngol 2001;110:142–147.
65 Harvey RJ, Parmar P, Sacks R, Zanation AM: Endoscopic skull base reconstruction of large dural defects: a systematic review of published evidence. Laryngoscope 2012;122:452–459.
66 Hubbard JL, McDonald TJ, Pearson BW, Laws ER Jr: Spontaneous cerebrospinal fluid rhinorrhea: evolving concepts in diagnosis and surgical management based on the Mayo Clinic experience from 1970 through 1981. Neurosurgery 1985;16:314–321.
67 Schlosser RJ, Woodworth BA, Wilensky EM, Grady MS, Bolger WE: Spontaneous cerebrospinal fluid leaks: a variant of benign intracranial hypertension. Ann Otol Rhinol Laryngol 2006;115:495–500.
68 Woodworth BA, Prince A, Chiu AG, et al: Spontaneous CSF leaks: A paradigm for definitive repair and management of intracranial hypertension. Otolaryngol Head Neck Surg 2008;138:715–720.
69 Woodworth BA, Palmer JN: Spontaneous cerebrospinal fluid leaks. Curr Opin Otolaryngol Head Neck Surg 2009;17:59–65.

Bradford A. Woodworth
Department of Surgery, Division of Otolaryngology, University of Alabama-Birmingham
BDB 563 1530 3rd Ave S
Birmingham, AL 35294–0012 (USA)
E-Mail bwoodwo@uab.edu

Bleier BS (ed): Comprehensive Techniques in CSF Leak Repair and Skull Base Reconstruction.
Adv Otorhinolaryngol. Basel, Karger, 2013, vol 74, pp 163–173 (DOI: 10.1159/000342293)

Prevention and Management of Complications after Radiotherapy for Skull Base Tumors: A Multidisciplinary Approach

Joanne W. Jang · Annie W. Chan

Department of Radiation Oncology, Massachusetts General Hospital, Boston, Mass., USA

Abstract

Radiotherapy is an important treatment approach for a variety of benign and malignant skull base tumors. For patients with malignant skull base tumors, radiation therapy is usually given with surgery and chemotherapy. While patient- and tumor-specific factors can affect the risk of radiation-related complications, surgery and chemotherapy also play a large causal role. As surgery and chemotherapy can lower the radiation dose tolerance of the normal tissues in the skull base, a multidisciplinary team approach is necessary in the initial treatment decision-making process. Radiation can delay wound healing and increase complications after reconstructive surgery, and so surgical planning between surgeons and radiation oncologists is critical to ensure the successful repair. In this article, we will discuss the etiology, risk factors and management of osteonecrosis and sinonasal-cutaneous fistula, the two major complications after multimodality treatment of skull base tumors. Careful consideration of specific tumor-, patient- and treatment-related factors in a multidisciplinary manner is vital in minimizing these complications.

Radiation therapy is an important treatment for patients with benign and malignant skull base tumors. Radiation therapy can be given alone or in conjunction with surgery or chemotherapy. If used with surgery, radiation therapy is usually given postoperatively, although on rare occasions, it can be given prior to skull base surgery.

Unfortunately, the dose that is required to control some of the aggressive skull base malignancies exceeds the radiation dose tolerance of the normal tissues in the skull base. Surgery and chemotherapy further lower the dose tolerance of normal tissues, and treatment-related side effects are often unavoidable after radiation therapy alone or in combinations with surgery or chemotherapy. In this review article, we review the two common side effects after surgery and radiation to the skull base, osteonecrosis and sinonasal-cutaneous fistula.

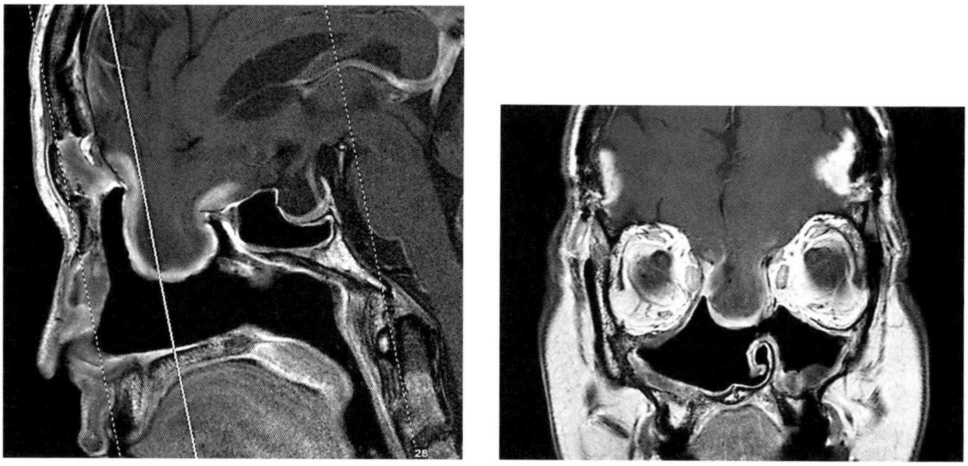

Fig. 1. 51-year-old female with a low-grade esthesioneuroblastoma of the right nasal cavity. She was treated with craniofacial resection and postoperative radiation therapy. Fifteen months after treatment and 3 months after a biopsy of the nasal roof, she experienced cerebrospinal fluid leakage. She visualized her brain tissue through her nasal cavities during irrigation. Imaging revealed meningoencephalocele through the flap reconstruction. She underwent reconstructive surgery with a fascia lata flap and has done well since.

Osteonecrosis

Etiology and Risk Factors

Osteonecrosis of the anterior, central and lateral skull base can occur after almost any type of treatment involving the skull base [1–3]. Unlike osteonecrosis of the mandible, osteonecrosis of the bones of the skull base is less well studied and reported. The bones that can be affected include the clivus, temporal bone and sphenoid sinus. The sequelae of osteonecrosis of the skull base can range from relatively mild to severe, and can result in fistula, cerebral herniation (fig. 1), cerebrospinal fluid (CSF) leak, pneumocephalus (fig. 2), meningitis, or death.

Patient-Related Factors
Certain patient-specific clinical factors can affect the vasculature in the skull base, which then can affect the risk of osteonecrosis. For example, diabetes, hypertension, tobacco smoking, peripheral vascular disease, atherosclerotic disease, or vasculitides can all result in decreased blood supply to the skull base, and in the setting of treatment with surgery, radiation or chemotherapy, can increase the risk of osteonecrosis.

Tumor-Related Factors
The location, extent, and type of tumor can also affect the risk of osteonecrosis. Larger tumors and tumors that involve the bones of the skull base are more likely to be associated with osteonecrosis. These tumors have likely weakened the skull

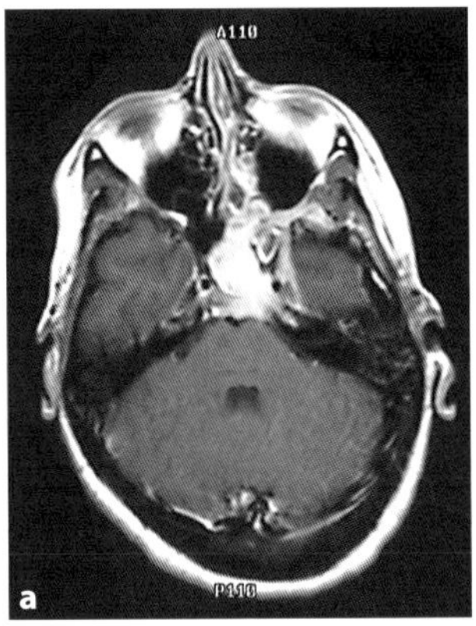

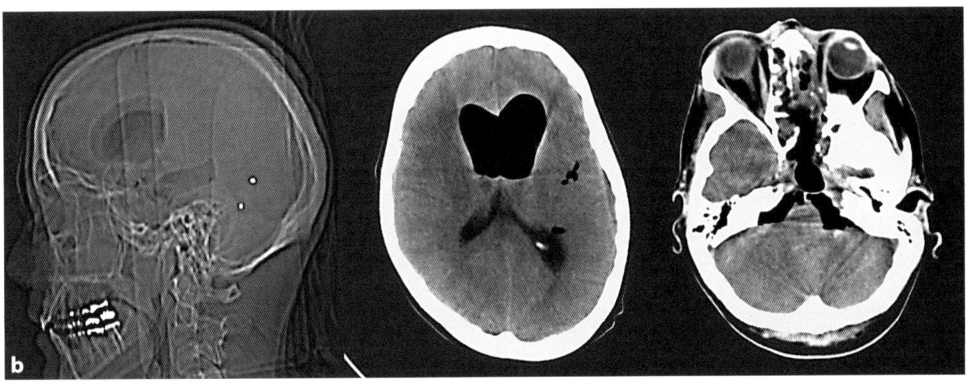

Fig. 2. **a** 46-year-old female with adenoid cystic carcinoma involving the sphenoid sinus, cavernous sinus, ethmoid sinus, nasopharynx, clivus, and prepontine cistern. The patient was treated with definitive radiation therapy. **b** Two months after completion of radiation, the patient developed a fistula between pontine cistern and the nasal cavity. She developed recurrent cerebrospinal fluid leaks and infectious meningitis. She later died due to extensive intraventricular and subarachnoid air.

base, or sometimes infiltrate the bones and become part of the supportive nature of the skull base. With tumor excision, or tumor regression with any type of treatment, the bones are left weaker, without even the support of the tumor, and become even more prone to osteonecrosis. For tumors with a propensity to regress rapidly during treatment with chemotherapy or radiation, and without time for adequate remodeling, the bones in the skull base can become weaker and undergo osteonecrosis (fig. 3).

Treatment-Related Factors

Surgery

The most well-known cause of osteonecrosis is radiation or so-called osteoradionecrosis (ORN), although the use of surgery either before or after radiotherapy can increase the risk of such an event. As achieving gross total resection is the goal of

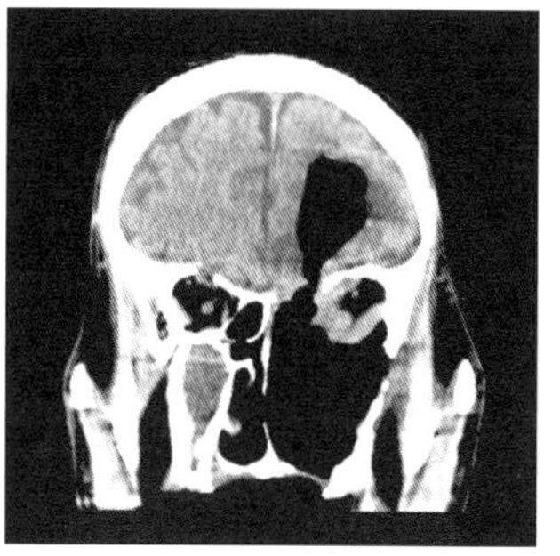

Fig. 3. A 46-year-old patient with recurrent nasopharyngeal carcinoma. The patient was treated with chemotherapy. He developed osteoradionecrosis as a result of the rapid good response to chemotherapy. He eventually died of pneumocephalus after a big sneeze.

skull base resection for most skull base malignancies, it is not uncommon for surgeons to resect the involved bones of the skull base at the time of radical surgery. Not only can this result in thinning and compromised integrity of the bones in the skull base, but it can decrease the vascular supply to the skull base, increasing the risk of ORN. Postoperative complications, such as infection, graft failure or damaged vascular supply of the skull base, all are factors that have been associated with ORN, and can increase the risk of ORN if patients are treated with radiation following surgery.

Chemotherapy

It is well known that chemotherapy sensitizes tumors to radiation, and the effect of chemotherapy on normal tissue sensitization is underappreciated. The addition of concurrent chemotherapy to radiation can also increase the risk of radiation-induced ORN. However, chemotherapy alone can cause osteonecrosis and damage to the skull base. Some tumors, such as nasopharyngeal carcinomas, are very chemotherapy-sensitive. Unlike radiation, for which tumor regression is relatively gradual providing time for bone remodeling and remineralization, during induction chemotherapy, these tumors can rapidly regress, without allowing time for reossification. As a result, there can be acute and extensive damage to the skull base, resulting in brain herniation, CSF leak, pneumocephalus, infection or death (fig. 3).

Radiation Therapy

There are also radiation-related factors that can influence the incidence of ORN. A higher total radiation dose, higher dose per fraction, and accelerated fractionation schema can increase the risk of ORN. Smaller fraction sizes, even if given twice a day rather than once daily, can also decrease the likelihood of ORN [4–6]. Although no threshold dose has been described for the base of skull, in general, the skull base can receive 70–76 Gy of radiation using 1.8–2.0 Gy fractions, without an unacceptably high risk of ORN, if the patient has no other underlying risk factors.

The size, location and heterogeneity of the radiation treatment plan can also affect the risk of ORN. ORN of these bones is related to whether or not these bones are included in the radiation treatment field, and the degree to which they are included. The larger the area of bone irradiated, the higher the probability of osteonecrosis. Due to the presence of air cavities and bones in the skull base, it is common to have dose inhomogeneity and therefore hot spots in the radiation treatment plans. If a hot spot unfortunately falls within a bone of the skull base, this could result in increased risk of ORN.

For some patients with recurrent disease, further surgery may not be an option, and re-irradiation to the skull base may be indicated. For these patients, development of ORN is expected, and patients should be treated only with the knowledge of this consequence.

Prevention and Management

Surgical Considerations

For patients receiving postoperative radiation to the skull base, every attempt should be made to minimize the risk of ORN. If a particular bone in the skull base has been severely damaged from tumor resection that area can be reinforced with a vascularized tissue flap to provide structural support and improve blood supply. This can actually be done at the time of skull base reconstructive surgery, or in some cases, as a second procedure before starting radiotherapy if the risk of ORN and its associated consequences are thought to be high enough.

For many malignancies, the time it takes to start radiation after surgery can affect survival outcomes and starting radiation as soon as possible after adequate healing is often a priority. The benefit of using a vascularized flap is faster and more adequate healing, and therefore, the earlier and safer delivery of radiation [7]. These include regional pedicled flaps, free flaps, and endoscopic reconstructions. In support of this idea, one study cited the use of a pedicled pericranial flap to prevent osteoradionecrosis if postoperative radiotherapy was planned, rather than their traditional method of using a double layer fascial graft for skull base reconstruction [8]. Similarly, others support the use of vascularized bone, or wrapping pericranium or grafts in vascularized tissue to mitigate the potential side effects of postoperative radiation [8–11]. After surgery, every effort must be made and the patient must be under careful observation to ensure there is no vascular compromise, infection or graft failure, to decrease the probability of ORN after radiation. Frequent endoscopic examinations may assist in monitoring the viability of the reconstruction thereby providing the opportunity to closely titrate the initiation of radiotherapy to the degree of wound healing.

In the less common setting of preoperative radiation, there can be radiation-induced changes in the vasculature and soft tissues. In these cases, grafts from nonirradiated tissue, such as a free flap from a distant donor site, may result in a lower risk of postoperative complications. For this reason, some groups advocate the use of free flaps and microvascular techniques for patients that have undergone prior radiotherapy [12–15]. Moreover, if possible, skull base reconstructive techniques should attempt to utilize nonirradiated recipient tissue and vasculature [9]. Suggestions for alternative vessels to use for anastomoses in free tissue transfer include the transverse cervical vessels, the thoracoacromial artery and cephalic vein, and the internal mammary vessels [16–21].

Nevertheless, there is no consensus as to which specific type of flap should be used when considering preoperative or postoperative radiotherapy. One investigation showed that there was no difference between pedicled and free flaps in postoperative complication rates after both reconstructive surgery and radiation [22]. No studies have compared regional pedicled flaps or free flaps to endoscopic flap skull base reconstructions, which have more recently come into widespread use. However, one institution has published several series using multiple types of endoscopic approaches for skull base reconstruction, with postoperative radiation one of the indications for such an approach [7, 23, 24]. These include endoscopic pedicled endonasal, pericranial, and temporoparietal flaps.

Radiotherapeutic Considerations

Specific radiation-related factors can also be modified to try and decrease the probability of development of ORN. However, some of these factors, such as radiation dose or fractionation, may also be related to the outcomes of survival and tumor control, and altering the factors for the purposes of sparing toxicities must be balanced with the likelihood of recurrence. Similarly, as severe osteoradionecrosis of the skull base could result in severe consequences such as death, the potential benefits and risks of re-irradiation should be carefully considered in individual cases.

Regardless of the treatment algorithm, radiation oncologists should always attempt to minimize the amount of skull base in the treatment portal, and to limit the dose to specific bones if possible, keeping in mind the need for adequate dose and radiation coverage of the tumor or surgical bed. In this era, all malignancies of the skull base are treated with intensity-modulated radiation therapy (IMRT) or proton beam, rather than conventional three-dimensional or two-dimensional radiotherapy. This allows providers to restrict doses to critical structures, including the sensitive bones of the skull base, in order to decrease the risk of ORN.

Patient Considerations

Radiation commonly results in chronic sinonasal congestion. This, in turn, can lead to increased risk of chronic nasal and sinus infections, which are often thought of as causal components of ORN. In order to try to prevent ORN of the skull base, it is essential to maintain good sinonasal hygiene. This includes routine endoscopic exams, cleanings, or debridement, daily nasal irrigation with or without antibiotics, and the cessation of tobacco abuse.

Treatment
Should osteoradionecrosis occur despite all preventative measures, treatment can range from conservative management with antibiotics for small lesions, to the need for extensive resection and reconstruction for large lesions. Hyperbaric oxygen has also been used by some institutions to treat osteoradionecrosis [25, 26]. However, the questionable efficacy, limited cost effectiveness, and relative inconvenience of hyperbaric oxygen have precluded its widespread use.

Sinonasal-Cutaneous Fistulas

Etiology and Risk Factors

The most common types of fistulas experienced after anterior skull base surgery, with or without radiation, are sinonasal-cutaneous fistulas. These fistulas can be a cosmetic problem and infection risk.

Patient-Related Factors
Similar to osteonecrosis, certain patient factors that affect vascularization of tissue can affect the risk of developing a sinonasal-cutaneous fistula. For example, diabetes, hypertension, smoking, vascular disease, atherosclerotic disease, vasculitides or trauma can decrease blood flow to normal tissues, and after an injury to the skin, can result in sinonasal-cutaneous fistula.

Tumor-Related Factors
Tumors that are large or located close to the cutaneous or subcutaneous tissues are more likely to be associated with development of sinonasal-cutaneous fistulas. After surgical resection or local radiation therapy, the skin in this area can be thinner, devascularized and of poor quality. As a result, any further injury or infection can easily disrupt the tissue and cause the formation of a sinonasal-cutaneous fistula.

Treatment-Related Factors
After radiation, the rate of sinonasal-cutaneous fistulas range from 1.2 to 10% [27]. These fistulas primarily occur in patients who have had prior surgery, particularly those that received transfacial incisions, and in cases where the tumor involved the subcutaneous tissues. The fistulas develop at the site of the incision or tumor involvement, and can be exacerbated by further trauma, infection

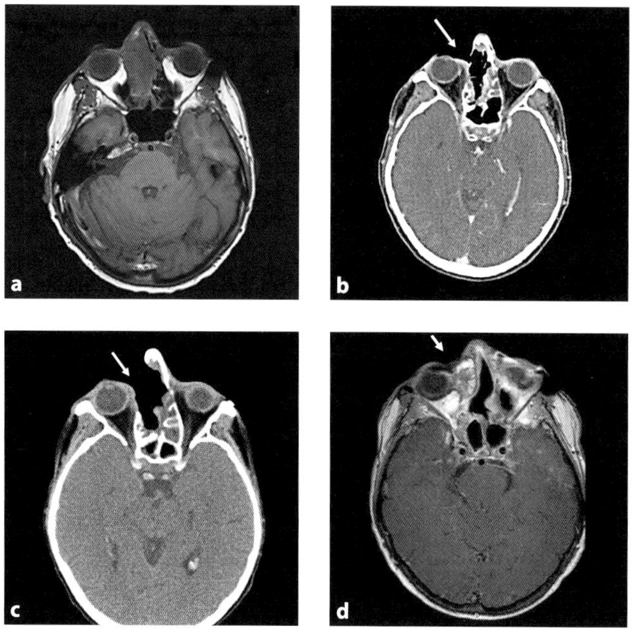

Fig. 4. **a** A 49-year-old with T4b squamous cell carcinoma of the nasal cavity. Preoperatively, the tumor extended to the preseptal orbital fat at the right medial canthus. He underwent craniotomy and right transfacial surgery. **b** Postoperatively, he was left with a nasocutaneous fistula, which was not corrected before starting concurrent chemoradiation. **c** After concurrent chemoradiation, the fistula enlarged. He underwent reconstructive surgery with rotational flap local tissue transfer, which resulted in further enlargement of the fistula. **d** The fistula was subsequently repaired with an anterior lateral thigh-free flap.

of the area, or radiation (fig. 4). Similar to ORN, a high dose of radiation or larger fractions of radiation, the addition of concurrent chemotherapy, and repeat irradiation to the same area can also increase the probability of sinonasal cutaneous fistulas.

Prevention and Management

Surgical Considerations

Sinonasal cutaneous fistulas are often an issue of reconstruction, and attempts can be made to minimize the probability of their development before starting radiation. As transfacial surgeries often remove bone and supportive tissues, leaving only a thin layer of skin, they can result in an increased likelihood of sinonasal cutaneous fistula after radiation when compared to endoscopic surgery. For patients with these malignancies, endoscopic surgeries, if possible, may result in a better cosmetic outcome and likely less need for further reconstructive surgeries after radiation. If there is already an existing wound, fistula or infection, radiation can make these problems worse, and it is better to perform immediate reconstructive surgery and allow adequate healing before starting radiation.

Radiotherapeutic Considerations
Specific radiation parameters can also be modified to decrease the likelihood of sinonasal cutaneous fistula. For example, attempts should be made to limit the dose of radiation to the subcutaneous soft tissues and the skin, decrease fraction sizes, and avoid re-irradiation. As there is increased heterogeneity of radiation dose distributions in or near the air cavities in the skull base, special care must be taken so that there will be no excess dose delivered to the skin or subcutaneous tissues. As for ORN, adjusting these factors must be balanced with the goal of tumor control.

Patient Considerations
As mentioned previously, radiation to the skull base can result in chronic sinus and nasal congestion, which can predispose patients to chronic infections. These infections play a causal role in the development of sinonasal cutaneous fistulas, and good sinonasal hygiene with daily irrigations and frequent examinations or debridement is necessary to try to prevent them.

Management
Should a sinonasal cutaneous fistula develop despite all attempts at prevention, it can be a serious reconstructive surgery issue. Further surgery using tissue in the immediate vicinity, which has often also received radiation, can be problematic as radiation can induce fibrosis in the soft tissues and vasculature. In fact, techniques using local pedicled flaps can result in development of further sinonasal cutaneous fistulas. Donor flap tissue should preferably come from distant sites that were not previously irradiated, such as the thigh or forearm, to ensure the highest likelihood of graft success.

Conclusion

ORN and sinonasal cutaneous fistulas are common complications after multidisciplinary treatment of the skull base. There are specific patient, surgical and radiation factors that can be considered and potentially modified to decrease the risk of such complications. The use of vascularized reconstructive techniques and endoscopic approaches may help to mitigate complications. Open and ongoing communication between surgeons, medical oncologists and radiation oncologists in a multidisciplinary manner is vital in order to prevent these complications and minimize the risk of ORN and sinonasal cutaneous fistulas.

References

1 Birzgalis AR, Ramsden RT, Farrington WT, Small M: Severe radionecrosis of the temporal bone. J Laryngol Otol 1993;107:183–187.

2 Huang XM, Zheng YQ, Zhang XM, Mai HQ, Zeng L, Liu X, Liu W, Zou H, Xu G: Diagnosis and management of skull base osteoradionecrosis after radiotherapy for nasopharyngeal carcinoma. Laryngoscope 2006;116:1626–1631.

3 Lim BY, Pang KP, Low WK, Tan HM: CSF otorrhea complicating temporal bone osteoradionecrosis in a patient with nasopharyngeal carcinoma. ENT J 2005;84:39–40.

4 Lee AW, Foo W, Chappell R, Fowler JF, Sze WM, Poon YF, Law SC, Ng SH, O SK, Tung SY, Lau WH, Ho JH: Effect of time, dose, and fractionation on temporal lobe necrosis following radiotherapy for nasopharyngeal carcinoma. Int J Radiat Oncol Biol Phys 1998;40:35–42.

5 Lee AW, Kwong DL, Leung SF, Tung SY, Sze WM, Sham JS, Teo PM, Leung TW, Wu PM, Chappell R, Peters LJ, Fowler JF: Factors affecting risk of symptomatic temporal lobe necrosis: significance of fractional dose and treatment time. Int J Radiat Oncol Biol Phys 2002;53:75–85.

6 Metson R, Freehling DJ, Wang CC: Surgical complications following twice-a-day versus once-a-day radiation therapy. Laryngoscope 1988;98:30–34.

7 Fortes FS, Carrau RL, Snyderman CH, Kassam A, Prevedello D, Vescan A, Mintz A, Gardner P: Transpterygoid transposition of a temporoparietal fascia flap: a new method for skull base reconstruction after endoscopic expanded endonasal approaches. Laryngoscope 2007;117:970–976.

8 Fliss DM, Gil Z, Spektor S, Leider-Trejo L, Abergel A, Khafif A, Amir A, Gur E, Cohen JT: Skull base reconstruction after anterior subcranial tumor resection. Neurosurg Focus 2002;12:e10.

9 Chim H, Salgado CJ, Seselgyte R, Wei FC, Mardini S: Principles of head and neck reconstruction: an algorithm to guide flap selection. Semin Plast Surg 2010;24:148–154.

10 Gil Z, Abergel A, Leider-Trejo L, Khafif A, Margalit N, Amir A, Gur E, Fliss DM: A comprehensive algorithm for anterior skull base reconstruction after oncological resections. Skull Base 2007;17:25–37.

11 Gok A, Erkutlu I, Alptekin M, Kanlikama M: Three-layer reconstruction with fascia lata and vascularized pericranium for anterior skull base defects. Acta Neurochir (Wien) 2004;146:53–56.

12 Chang DW, Langstein HN, Gupta A, De Monte F, Do KA, Wang X, Robb G: Reconstructive management of cranial base defects after tumor ablation. Plast Reconstr Surg 2001;107:1346–1355.

13 Georgantopoulou A, Hodgkinson PD, Gerber CJ: Cranial-base surgery: a reconstructive algorithm. Br J Plast Surg 2003;56:10–13.

14 Hanasono MM, Silva A, Skoracki RJ, Gidley PW, DeMonte F, Hanna EY, Chang DW, Yu P: Skull base reconstruction: an updated approach. Plast Reconstr Surg 2011;128:675–686.

15 Newman J, O'Malley BW Jr, Chalian A, Brown MT: Microvascular reconstruction of cranial base defects: an evaluation of complication and survival rates to justify the use of this repair. Arch Otolaryngol Head Neck Surg 2006;132:381–384.

16 Harris JR, Lueg E, Genden E, Urken ML: The thoracoacromial/cephalic vascular system for microvascular anastomoses in the vessel-depleted neck. Arch Otolaryngol Head Neck Surg 2002;128: 319–323.

17 Horng SY, Chen MT: Reversed cephalic vein: a lifeboat in head and neck free-flap reconstruction. Plast Reconstr Surg 1993;92:752–753.

18 Jacobson AS, Eloy JA, Park E, Roman B, Genden EM: Vessel-depleted neck: techniques for achieving microvascular reconstruction. Head Neck 2008;30: 201–207.

19 Schneider DS, McClain L, Robb PK Jr, Rosenthal EL, Wax MK: Use of internal mammary vessels in head and neck microvascular reconstruction. Arch Otolaryngol Head Neck Surg 2012;138:172–176.

20 Urken ML, Higgins KM, Lee B, Vickery C: Internal mammary artery and vein: recipient vessels for free tissue transfer to the head and neck in the vessel-depleted neck. Head Neck 2006;28:797–801.

21 Yu P: The transverse cervical vessels as recipient vessels for previously treated head and neck cancer patients. Plast Reconstr Surg 2005;115:1253–1258.

22 Wang Z, Qiu W, Mendenhall WM: Influence of radiation therapy on reconstructive flaps after radical resection of head and neck cancer. Int J Oral Maxillofac Surg 2003;32:35–38.

23 Kassam AB, Thomas A, Carrau RL, Snyderman CH, Vescan A, Prevedello D, Mintz A, Gardner P: Endoscopic reconstruction of the cranial base using a pedicled nasoseptal flap. Neurosurgery 2008;63: ONS44–52.

24 Zanation AM, Carrau RL, Snyderman CH, Germanwala AV, Gardner PA, Prevedello DM, Kassam AB: Nasoseptal flap reconstruction of high flow intraoperative cerebral spinal fluid leaks during endoscopic skull base surgery. Am J Rhinol Allergy 2009;23:518–521.
25 Peterson DE, Doerr W, Hovan A, Pinto A, Saunders D, Elting LS, Spijkervet FK, Brennan MT: Osteoradionecrosis in cancer patients: the evidence base for treatment-dependent frequency, current management strategies, and future studies. Support Care Cancer 2010;18:1089–1098.
26 Spiegelberg L, Djasim UM, van Neck HW, Wolvius EB, van der Wal KG: Hyperbaric oxygen therapy in the management of radiation-induced injury in the head and neck region: a review of the literature. J Oral Maxillofac Surg 2010;68:1732–1739.
27 Cianchetti M, Varvares MA, Deschler DG, Liebsch NJ, Wang JJ, Chan AW: Risk of sinonasal-cutaneous fistula after treatment for advanced sinonasal cancer. J Surg Oncol 2012;105:261–265.

Annie W. Chan, MD
Department of Radiation Oncology, Massachusetts General Hospital
55 Fruit Street, Cox 337
Boston, MA 02114 (USA)
E-Mail awchan@partners.org

Bleier BS (ed): Comprehensive Techniques in CSF Leak Repair and Skull Base Reconstruction.
Adv Otorhinolaryngol. Basel, Karger, 2013, vol 74, pp 174–183 (DOI: 10.1159/000342294)

Novel Techniques and the Future of Skull Base Reconstruction

Joshua C. Meier · Benjamin S. Bleier

Department of Otology and Laryngology, Massachusetts Eye and Ear Infirmary, Harvard Medical School, Boston, Mass., USA

Abstract

The field of endoscopic skull base surgery has evolved considerably in recent years fueled largely by advances in both imaging and instrumentation. While the indications for these approaches continue to be extended, the ability to reconstruct the resultant defects has emerged as a rate-limiting obstacle. Postoperative failures with current multilayer grafting techniques remain significant and may increase as the indications for endoscopic resections continue to expand. Laser tissue welding represents a novel method of wound repair in which laser energy is applied to a chromophore doped biologic solder at the wound edge to create a laser weld (fig. 1). These repairs are capable of withstanding forces far exceeding those exerted by intracranial pressure with negligible collateral thermal tissue injury. Recent clinical trials have demonstrated the safety and feasibility of endoscopic laser welding while exposing the limitations of first generation hyaluronic acid based solders. Novel supersaturated gel based solders are currently being tested in clinical trials and appear to possess significantly improved viscoelastic properties. While laser tissue welding remains an experimental technique, continued success with these novel solder formulations may catalyze the widespread adoption of this technique for skull base repair in the near future.

The field of endoscopic skull base surgery has evolved significantly in recent years fueled largely by advances in both imaging and instrumentation. While the indications for these approaches continue to be extended, the ability to reconstruct the resultant defects has emerged as a rate-limiting obstacle. Successful reconstruction requires a technique that is able to recapitulate the complex morphology of the cranial base while simultaneously withstanding the hydrostatic forces exerted by the cerebrospinal fluid (CSF) compartment. Current techniques have evolved to compensate for the inherent restrictions of endoscopic access and thus represent a departure from the traditional principles of surgical closure in an open field. In place of meticulous tissue apposition and watertight suture lines, surgeons are currently limited to multilayer grafting and the liberal application of relatively weak biologic glues. These adhesives often obscure the defect thereby impairing the intraoperative assessment of the closure. Furthermore, these techniques may require prolonged nasal packing and lumbar drain placement to further support the repair in the immediate postoperative period. Despite these limitations, the relatively low pressure head exerted by CSF coupled with the dense vascularity of the dura and mucosa allow for robust wound healing leading to successful closure in the vast majority of cases. The more recent introduction of vascularized rotational flaps has further improved outcomes in several series. However, there remains an overall 10% failure rate that may increase as skull base approaches continue to expand in the future. These limitations have catalyzed investigation into a repair method that can be applied endoscopically and is capable of producing instant, watertight tissue bonds. Laser tissue welding (LTW) has been advanced as one such technology that may satisfy these criteria and further reduce the morbidity and failure rate of cranial base reconstruction.

History of Laser Tissue Welding

Since their invention in the late 1950s, lasers have been successfully utilized in a variety of medical fields as both diagnostic and therapeutic instruments. Lasers deliver a spatially and temporally coherent band of light at a fixed wavelength. These attributes are useful in medical applications as the narrow, focused beam allows for the controlled delivery of energy to the desired target [1]. Lasers were first used clinically in ophthalmologic applications; however, their use rapidly disseminated to other fields [2]. As early as 1960, the potential for use of lasers in tissue adhesion was recognized. In 1962, Sigel and Acevado [3] were the first to report the use of thermal energy in tissue adhesion by utilizing high frequency electric current to create an end-to-side porto-caval shunt in a canine model. Yahr and Strully [4] adapted this concept to light energy and are credited with the first description of LTW for vascular anastomosis. In 1979, Jain and Gordisch [5] used a neodymium:yttrium aluminum garnet (Nd:YAG) laser to repair incisions made in rat carotid and femoral arteries and later refined the technology to create rat carotid to carotid anastomoses as small as 0.6 mm [6]. Jain went on to use this technology in humans, using an Nd:YAG laser to successfully perform superficial temporal to middle cerebral artery bypasses in 5 patients. The total operating time was reported at 5 min compared to the 15 min needed for conventional hand-sewn anastomoses. At 6–9 months' follow-up, all vessels were reported to be angiographically patent [7]. Experiments in the 1980s also attempted to apply the CO_2 laser to vascular repairs. Unlike the Nd:YAG data, however, these studies found that the CO_2 energy led to aneurysm formation at the site of repair. Bass et al. [8] suggested that the avid absorption of the CO_2 laser by water resulted in suboptimal tissue penetration leading to overheating of the outer layers and minimal delivery of energy to the deeper tissues. These findings showed that the choice of an appropriate laser wavelength with its attendant absorption characteristics is critical to the success of the repair. As technology advanced, a wider variety of laser systems were introduced into the clinical armamentarium including the argon and diode lasers which allowed researchers to further tailor the wavelength and depth of penetration to their tissue of interest [2]. Despite this, in vivo reports of LTW using laser energy alone continued to be limited by suboptimal wound tensile strength and complications related to thermal leakage to surrounding healthy tissues.

Biologic Solders and Chromophores

A critical step in overcoming these limitations came over the last three decades with the introduction of biologic solder materials coupled with wavelength specific chromophores [2, 9]. These materials allow for improved target specific laser energy absorption with enhanced bond strengths and a reduction in the degree of collateral thermal leakage [10]. As early as 1979, Jain and Gorisch [5] used blood as a de facto solder and acknowledged that 'the minute quantity of blood helps in the absorption of the laser beam'. Multiple subsequent experiments have been performed using a variety of solder materials including blood, fibrin, albumin and cryoprecipitate. These have consistently demonstrated increased strength, decreased rate of stricture, and decreased formation of vascular aneurysms when compared with laser energy alone [11]. Talmor et al. [2] compared rat urethral repair with suture, CO_2 laser, and CO_2 laser with a protein solder. Their success rates were 50, 58 and 90%, respectively. Further experiments carried out by Wright and Poppas demonstrated that increasing the albumin content of the solder from 25 to 50% improved wound strength [12]. Histologic analysis of wounds repaired using a biologic solder demonstrated that the direct thermal injury was restricted to the region immediately adjacent to the solder without evidence of underlying tissue injury [10, 13]. This is further supported by a study in an esophageal fistula model in which no thermal injury was noted in the esophageal mucosa following laser-assisted closure using an albumin based solder [14].

Working off the success of these protein-based solutions, investigators sought to further enhance the efficiency of solder specific energy absorption by doping them with chromophores. Chromophores are molecules that preferentially absorb a specific wavelength of visible electromagnetic radiation. The addition of chromophores provides a method of concentrating the laser energy within the solder enabling adequate lasing with lower irradiances and reduced dispersion of heat to surrounding tissues. Previously studied chromophores include indocyanine green (ICG), fluorescein, and carbon black. Of these, ICG has emerged as the most widely studied chromophore used in LTW with a peak absorption wavelength of 805 nm [2]. This chromophore

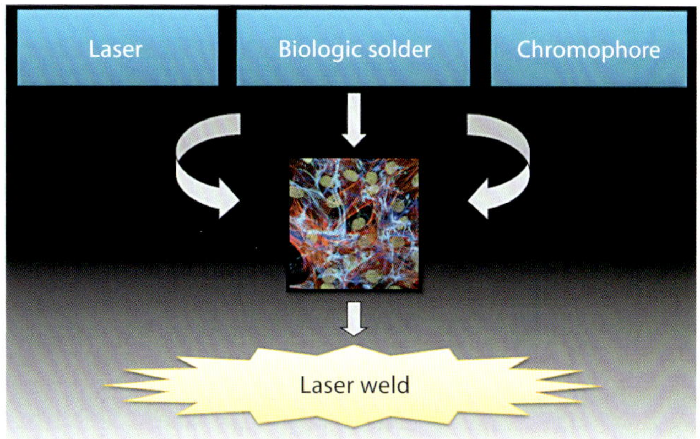

Fig. 1. Components required to create a successful laser weld when applied to a wound.

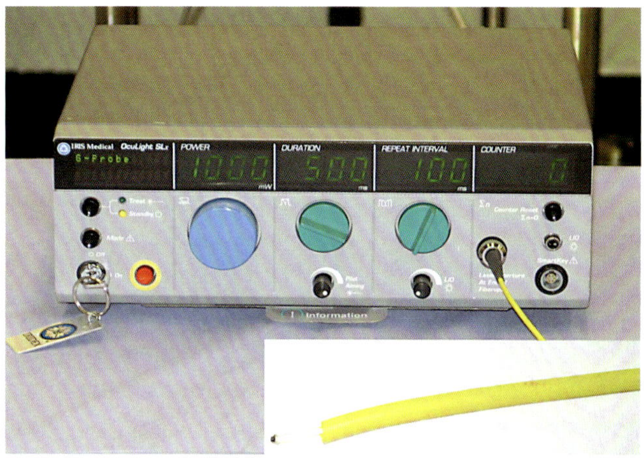

Fig. 2. 810-nm diode laser module with its associated quartz-silica fiber optic cable (inset) used to deliver the laser energy.

is targeted by the 808-nm diode laser which has been utilized in a variety of animal and clinical studies and is minimally absorbed by most native tissues (fig. 2) [14]. The chromophore also provides a secondary benefit by providing an objective basis of gauging adequacy of lasing through a predictable color change [15]. Some authors have described a computer-based thermal feedback system to further define the lasing endpoint [10]; however, its practical utility in light of the negligible thermal effect previously described remains to be seen.

Mechanism of Laser Tissue Welding

While a considerable amount of research has focused on optimizing LTW techniques in a variety of animal models, the precise mechanism of tissue bonding has yet to be fully explained. Most authors agree that the application of laser energy results in a restructuring of the extracellular matrix with resultant interactions of native tissue proteins including collagen interdigitation, and noncovalent laminin and entactin bonding [13, 16, 17]. In 1986, Schober et al. [18] welded rat carotid arteries with a 1.32-μm Nd:YAG laser. By performing electron microscopy on the arterial welds they found that the laser loosened the collagen triple helix leading to rearrangement and interdigitation between neighboring collagen fibers. White et al. [19] similarly performed electron microscopy on canine femoral anastomoses following argon laser welding and reported direct collagen to collagen and collagen to elastin bonds. Following CO_2 lasing of rabbit dura using an egg white solder, Menovksy et al. [20] found collagen to collagen bonding, changes in the collagen ultrastructure, and bonding of the solder directly to the collagen. Using a diode laser to repair rat aortotomies, Tang et al. [21] also reported collagen fusion, intertwining of opposing collagen fibrils to form 'ropes', and fibril swelling. Based on this they agreed that morphological changes of collagen are important for the mechanism of welding.

In contrast to these studies, Helmsworth et al. [17] applied the argon laser to extracellular basement membranes and suggested that while the laser altered the basement membrane protein structure it was not found to result the formation of new protein complexes. Rather, this group postulated the alteration of disulfide bonds as a potential mechanism due to the preferential absorption of the argon laser wavelength by these bonds. Murray et al. [22] demonstrated a decrease in both a 235-kDa guanidine extractable protein and type VI collagen with a concomitant rise in protein aggregates with noncollagenous domains suggesting that fibronectin is also involved. Bass et al. [8] lased rat tail tendons, which are composed of mostly type I collagen, using a diode laser and reported that while type I collagen was denatured no new covalent bonds were formed. They based this finding on a lack of high molecular weight complexes on electrophoresis. This group went on to examine multiple different LTW configurations and concluded that while covalent bonding may occur during LTW, it is not necessary for weld strength [11]. Although these studies have contributed to our grasp of the mechanism of laser tissue welding, a complete understanding remains confounded by the use of varying lasers, lasing times, solders, doping agents, and tissue substrates. Combined with inherent operator variation, these variables have led to a diverse set of experimental controls resulting in sometimes contradictory conclusions.

Wound Repair Strength and Collateral Injury with LTW

To date, LTW has been studied in a wide variety of tissues including blood vessels, gut, nerves, skin, dura, bladder, and urethra [23]. Despite our incomplete understanding of the mechanism, these studies have consistently confirmed that it is an efficacious method of tissue bonding capable of creating instant welds with significant tensile strength and high leak pressures. Barrieras et al. [23] compared laser welded pyeloplasty using an albumin/fibrin solder to both traditional suture and fibrin glue repair. This study found an immediate increase in anastomotic leak pressure utilizing a laser weld over both suture and fibrin glue (37.2 ± 1.1 mm Hg as compared to 12.8 ± 4.0 and 2.6 ± 1.1 mm Hg, respectively). In 1996, Foyt et al. [13] performed one of the first ex vivo studies examining the utility of laser tissue welding in a head and neck application. This group looked at primary human cadaveric dural closure utilizing an albumin/indocyanine green dye mixture. They reported a leak pressure of 26.2 ± 3.7 mm Hg with the laser closure as compared to 9.4 ± 1.7 mm Hg in the suture group. These studies underscore the fact that the efficacy of the weld depends on both the laser/solder combination as well as the tissue being addressed. The spectral absorption, extracellular matrix composition, and tensile properties of the underlying tissue all play an important role in the behavior of the lased coagulum.

The 808-nm diode laser coupled with a 42% human albumin/indocyanine green/hyaluronic acid-based solder has been the most extensively studied in both animal and clinical trials and has become the basis for investigation for the role of LTW in skull base reconstruction. In early studies examining LTW in tissues relevant to the skull base repair, burst strengths in sheep septal mucosa and periosteum were found to be significantly higher than those of suture repair (mucosa 34.88 ± 3.49 vs. 17.06 ± 1.83 mm Hg, $p = 0.0001$; periosteum 30.02 ± 2.23 vs. 24.12 ± 0.71 mm Hg, $p < 0.0001$) [24]. This data was used to establish an animal model in the New Zealand White rabbit to analyze the prospective behavior of these welds in the repair of dorsal maxillary sinusotomies [15]. This study found that laser welding was capable of achieving an immediate burst strength over four times that of normal human intracranial pressure. The strength of these bonds increased over the first 2 weeks and was noted to converge with those of an open wound left to heal by secondary intention by postoperative day 15 (fig. 3). This offers an interesting insight into the natural history of these wounds in a clinical setting where it appears that native scarring and fibrosis alone are capable of sealing these defects within several weeks confirming that efforts should be focused on preventing early repair failures. The other key finding of this study was that no thermal injury was noted in the surrounding tissues or underlying sinus mucosa (fig. 4). These findings are consistent with those of other investigators. Foyt et al. [13] described using a diode laser coupled to an ICG doped albumin-based solder to weld rat dura. They reported no thermal damage to the underlying cerebral cortex in eight out of ten samples. The lack of collateral thermal injury represents a critical attribute of LTW particularly when considering the proximity of these repairs to critical neurovascular structures in the skull base.

Prospective wound healing has also been examined in several studies and in contrast to the granulomatous inflammatory response seen with suture material, the lased solder provides a nonimmunogenic scaffold upon which fibroblast migration and scar formation may proceed [16]. This coagulum is gradually absorbed during

Fig. 3. Prospective comparison of burst strengths of rabbit maxillary sinus mucosal repairs using laser welding (LTW) vs. wounds left to heal by secondary intention (Control).

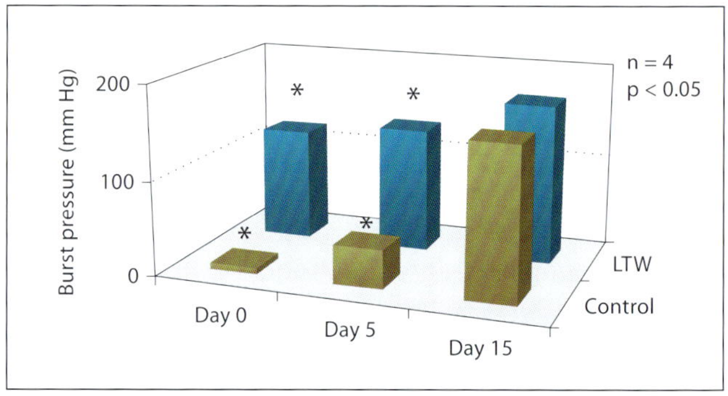

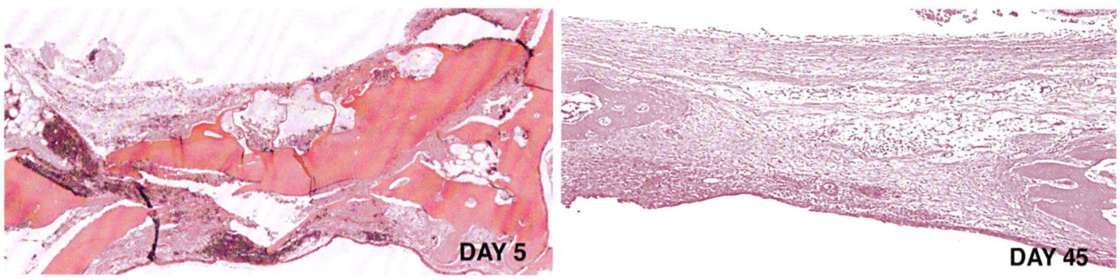

Fig. 4. Hematoxylin and eosin images of the rabbit maxillary sinusotomy following laser welding. On day 5, the eosinophilic solder is present bridging the wound with no surrounding thermal injury. By day 45, the solder has been completely absorbed leaving only an organized scar.

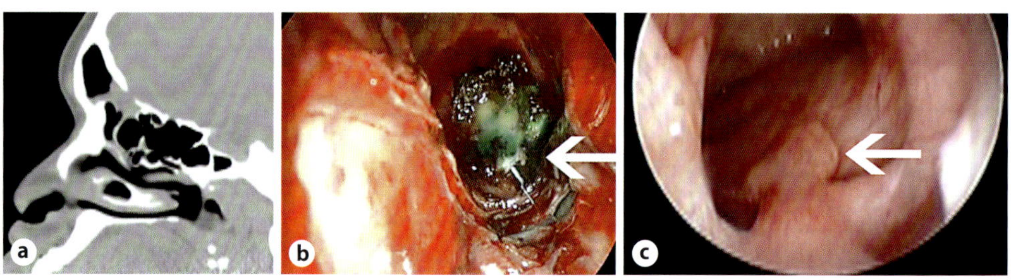

Fig. 5. **a** Sagittal CT scan demonstrating a left frontoethmoidal encephalocele. **b** Intraoperative view following laser welding of an underlay bone graft. **c** Endoscopic view of the repair on postoperative day 72 demonstrating a well-mucosalized defect with no residual solder.

the normal wound healing process [25, 26] with minimal disruption of elastic fibers [27] and a similar collagen density seen with traditional wound closure techniques [23]. Another study in porcine dura reported integration of vascular and fibrous tissue into the repair within a week confirming that normal wound healing is unimpeded by the solder enhanced weld [28].

Clinical Challenges in LTW for Endoscopic Skull Base Reconstruction

The preclinical studies described suggest that LTW is able to produce instant wound repairs capable of withstanding forces far exceeding those generated by intracranial pressure with negligible collateral thermal injury. Despite the success of these animal studies, clinical data is still required to determine the safety and technical feasibility of this technique in an operative field. Kirsch et al. [16] reported a clinical trial in open human ureteroplasty and noted significantly higher intraoperative leak threshold pressures in laser solder reinforced microsutured incision lines than in suture alone (94.2 ± 24.2 vs. 20.0 ± 2.9 mm Hg, $p < 0.001$) in a study of 10 patients. While this data is encouraging, there are unique challenges to the endoscopic application of this technique which must be overcome.

The fact that both the chromophore doped solder and laser energy can be applied through an endoscopic port suggests that this technology may be utilized to vastly improve the repair of endoscopic skull base defects. While weld strength is the primary determinant of the utility of LTW, limited data exists on the prospective effects on wound healing as well as the risk of thermal injury when applied endoscopically. This potential catalyzed a safety and feasibility study of the use of LTW in sinonasal mucosa using an albumin based solder containing ICG and hyaluronic acid. In 2010, Bleier et al. [29] compared endoscopically welded respiratory mucosal defects in human nasal cavities to control wounds left to heal by secondary intention. Independent blinded investigators scored both the LTW and control areas and no difference was found between the two groups with respect to inflammation, thermal injury or edema. This pilot study showed LTW to be both safe and technically feasible when performed endoscopically with a similar healing pattern to the control wounds. Despite these promising experimental results, the first-generation hyaluronic acid-based solder was found to have significant technical problems when applied clinically. As a liquid, the solder was unable to bridge tissue gaps and was subject to gravitational displacement. These findings suggest that the clinical success of endoscopic laser welding is highly dependent on the viscoelastic properties of the solder itself. In a low viscosity formulation such as the hyaluronic acid solution, the solder tends to run off the skull base towards the dependent areas of the field and can be diluted by blood and irrigation. These formulations also have a high water content thereby requiring a longer lasing time to heat the solution to the temperature required for protein denaturation. This is not only inconvenient but also prolongs the potential for thermal leakage to the surrounding tissue. Conversely, high viscosity solutions and solids may be difficult to apply endoscopically and are less likely to interpolate themselves into the interstices of the wound resulting in gaps, which can lead to repair failure.

In an effort to address these challenges, novel solder formulations are currently being developed which consist of a supersaturated gel capable of bridging gaps between wound edges and providing enhanced lasing speed and strength due to the reduced intrinsic water content. Early clinical experience with this new formulation in two spontaneous encephaloceles with associated CSF leak demonstrated immediate intraoperative sealing of the defect with no long term failures (fig. 5) [unpubl. data] (online suppl. video 1).

Conclusions

Reconstruction of the cranial base remains a significant challenge to the progress of endoscopic skull base surgery. Laser tissue welding offers the potential of creating instant, water tight repairs with negligible thermal tissue injury. Recent clinical trials have demonstrated the safety and feasibility of endoscopic laser welding while exposing the limitations of first-generation hyaluronic acid-based solders. Novel supersaturated gel-based solders are currently being tested in both animal and clinical trials and appear to possess significantly improved viscoelastic properties. While LTW remains an experimental technique, its future in the field of endoscopic skull base surgery remains bright.

References

1 Garrett CG, Ossoff RH, Reinisch L: Laser surgery: basic principles and safety considerations; in: Cummings Otolaryngology – Head and Neck Surgery. Philadelphia, Mosby, 2010.
2 Talmor M, Bleustein CB, Poppas DP: Laser tissue welding: a biotechnological advance for the future. Arch Facial Plast Surg 2001;3:207–213.
3 Sigel B, Acevado FJ: Vein anastomosis by electrocoaptive union. Surg Forum 1962;13:233–235.
4 Yahr WZ, Strully KJ: Blood vessel anastamosis by laser and other biomedical applications. J Assoc Adv Med Instrum 1966;1:28–31.
5 Jain KK, Gorisch W: Repair of small blood vessels with the neodymium-YAG laser: a preliminary report. Surgery 1979;85:684–688.
6 Jain KK: Sutureless microvascular anastomoses using a neodymium-YAG laser. J Microsurg 1980;1:436–439.
7 Jain KK: Sutureless extra-intracranial anastomosis by laser. Lancet 1984;324:816–817.
8 Bass LS, Moazami N, Pocsidio J, Oz MC, Logerfo P, Treat MR: Changes in type I collagen following laser welding. Lasers Surg Med 1992;12:500–505.
9 Oz MC, Johnson JP, Parangi S, et al: Tissue soldering by use of indocyanine green dye-enhanced fibrinogen with the near infrared diode laser. J Vasc Surg 1990;11:718–25.
10 Gil Z, Shaham A, Vasilyev T, et al: Novel laser tissue-soldering technique for dural reconstruction. J Neurosurg 2005;103:87–91.
11 Bass LS, Treat MR: Laser tissue welding: a comprehensive review of current and future clinical applications. Lasers Surg Med 1995;17:315–349.
12 Wright EJ, Poppas DP: Effect of laser wavelength and protein solder concentration on acute tissue repair using laser welding: initial results in a canine ureter model. Tech Urol 1997;3:176–181.
13 Foyt D, Johnson JP, Kirsch AJ, et al: Dural closure with laser tissue welding. Otolaryngol Head Neck Surg 1996;115:513–518.
14 Bleier BS, Gratton MA, Leibowitz JM, et al: Laser welded endoscopic endoluminal repair of iatrogenic esophageal perforation: an animal model. Otolaryngol Head Neck Surg 2008;139:713–717.
15 Bleier BS, Palmer JN, Gratton MA, et al: Laser tissue welding in the rabbit maxillary sinus. Am J Rhinol 2008;22:625–628.
16 Kirsch AJ, Miller MI, Hensle TW, et al: Laser tissue soldering in urinary tract reconstruction: first human experience. Urology 1995;46:261–266.
17 Helmsworth TF, Wright CB, Scheffter SM, et al: Molecular surgery of the basement membrane by the argon laser. Lasers Surg Med 1990;10:576–583.

18 Schober R, Ulrich F, Sander T, Durselen H, Hessel S: Laser-induced alteration of collagen substructure allows microsurgical tissue welding. Science 1986; 232:1421–1422.
19 White RA, Kopchok GE, Donaye CE, Peng SK, Fujitani RM, White GH, Uitto J: Mechanism of tissue fusion in argon laser-welded vein-artery anastomoses. Lasers Surg Med 1988;8:83–89.
20 Menovsky T, Beek JF, van Gemert MJC: Laser tissue welding of dura mater and peripheral nerves: a scanning electron microscopy study. Lasers Surg Med 1996;19:152–158.
21 Tang J, Godlewski G, Rouy S, Delacretaz G: Morphologic changes in collagen fibers after 830 nm diode laser welding. Lasers Surg Med 1997;21: 438–443.
22 Murray LW, Su L, Kopchok GE, et al: Crosslinking of extracellular matrix proteins: a preliminary report on a possible mechanism of argon laser welding. Lasers Surg Med 1989;9:490–496.
23 Barrieras D, Reddy PP, McLorie GA, et al: Lessons learned from laser tissue soldering and fibrin glue pyeloplasty in an in vivo porcine model. J Urol 2000;164:1106–1110.
24 Bleier BS, Palmer JN, Sparano AM, et al: Laser assisted cerebrospinal fluid leak repair: an animal model to test feasibility. Otolaryngol Head Neck Surg 2007;137:810–814.
25 Lauto A, Kerman I, Ohebshalon M, et al: Two-layer film as a laser soldering biomaterial. Lasers Surg Med 1999;25:250–256.
26 Lauto A, Trickett R, Malik R, et al: Laser-activated solid protein bands for peripheral nerve repair: an in vivo study. Lasers Surg Med 1997;21:134–141.
27 Abergel RP, Lyons R, Dwyer R, et al: Use of lasers for closure of cutaneous wounds: experience with Nd:YAG, argon and CO_2 lasers. J Dermatol Surg Oncol 1986;12:1181–1185.
28 Tachibana E, Saito K, Fukuta K, et al: Evaluation of the healing process after dural reconstruction achieved using a free fascial graft. J Neurosurg 2002;96:280–286.
29 Bleier BS, Cohen NA, Chiu AG, O'Malley BW, Doghramji L, Palmer JN: Endonasal laser tissue welding: first human experience. Am J Rhinol 2010;24:244–246.

Benjamin S. Bleier, MD
Department of Otology and Laryngology, Massachusetts Eye and Ear Infirmary
Harvard Medical School
243 Charles St., Boston, MA 02114 (USA)
E-Mail bleierb@gmail.com

Author Index

Subject Index